THE WATCH

from its origins to the XIXth century

Catherine Cardinal

THE WATCH
from its origins to the XIXth century

Translated by Jacques Pages

The dimensions of the watches reproduced here are given in millimeters.
Abbreviations: L: length; W: width; D: diameter; Th: thickness.

Tabard Press
27 West 20th Street
New York, NY 10011

Tabard Press is an imprint of
William S. Konecky Associates

ISBN: 0-914427-29-6

Printed and bound in Hong Kong

Contents

Introduction

Among the thought-provoking subjects the history of civilization offers, the measurement of time is certainly one of those that force the historian to approach the most diversified domains. Indeed, it is inextricably linked to the history of human societies, to technical developments and to decorative arts. Timepieces show evidence of the evolution of man's manners and customs and are both functional objects that mirror the technical progress of a country and objets d'art attesting to the renewal of ornamental styles.

The different developmental phases of the horological industry can be brought to light through the examination of social, economic and political conditions. The study of watch production leads one to an analysis of its guild organizations in different watchmaking cities in Europe. This analysis is the more indispensable in that guild regulations, along with governmental economic policies, often determined the success or the failure of the horological centers. The problems of competition should not be neglected, because at times they played an important role.

A study of these diverse facts permits us to distinguish the successive stages in the expansion of the horology "in miniature" from its beginnings to the early nineteenth century: from 1500 to 1600, the demand by a rich clientele favored the creation of the first watch centers, some as ephemeral as Augsburg, others destined for a bright future, such as Paris and Blois; from 1600 to 1660, the prestige of French watchmaking asserted itself, while Geneva and London were developing into major centers; from 1660 to 1770, the decline of the industry in France, partly caused by Huguenot emigration, contrasted with the vitality of its English counterpart and the soaring of the Genevan Factory; from 1730 to the start of the nineteenth century, a dazzling revival of French watchmaking occurred, counterbalanced by the growing prosperity of the Swiss industry.

The technical expansion of horology "in miniature" faithfully mirrors the scientific explosion of modern times. After taking advantage of the research and discoveries of the Renaissance, which permitted its creation, the watch movement kept its primitive appearance until seventeenth-century scientists devoted their attention to its problems. The discovery in 1675 by C. Huygens of the balance-regulating spiral spring made the watch truly a timekeeper and in so doing started a race by many different researchers to perfect its exactitude, led notably by Julien Le Roy, Thomas Tompion, and George Graham, which lasted until the mid-nineteenth century. Thanks to the work done by English and French watchmakers on marine chronometers and to the inventions of J. A. Lépine, later followed by A. L. Breguet, a decisive stage in the conquest of precision was reached in the last decades of the eighteenth century.

The history of watch decoration is not only determined by the constant change in decorative styles. It is also sometimes influenced by the important changes undergone by the watch movements over the centuries and by the adaptation of new decorative techniques, such as painting on enamel, which presented enamelists with a novel way to ornament cases. Further, the evolution of case designs was diversely varied by the manner in which the clientele of each country would impose upon its watchmakers the decorative forms to which it was most receptive. Thus we see the appearance in the eighteenth century of the "Turkish watches" directed to the Turkish market and "Chinese watches" meant for China.

The study of the successive changes in case design throws into higher relief these diverse factors in its evolution. From 1520 to 1675, when horological techniques are almost the same all over Europe, it is remarkable that the fashions governing watch decoration are international. As a consequence of the extreme diversity in the decorative techniques which are then applied, and the equally large variety of watchcase shapes, these fashions vary greatly: rock-crystal watches, oval or octagonal watches engraved in the manner of Delaulne, watches shaped like animals, flowers or common objects, watches decorated with champlevé enamels, watches ornamented with paint on enamel. From 1675 till about 1720, while important technical changes are achieved and the English watch, given a double case, diverges more and more clearly from the French "*oignon*" watch, national styles do appear. During this same period, however, the cases painted by the Huauds achieve an international vogue. From 1730 until the early years of the nineteenth century, the French watch and the English watch maintain their distinctiveness and reflect, each in its own way, the different decorative styles of the eighteenth century.

What part, through its historical evolution, its techniques and its decoratives styles, does small-work horology play, up to the nineteenth century, in the social and economic life and the artistic creation in Europe? That is the question this study will try to answer.

Our research was made easier, on the one hand, by high-quality studies, each analyzing the development of an important watchmaking center. Thus, the expansion of the industry in Blois was studied in detail by Abbé Develle in *Les Horlogers blésois aux XVI*[e] *et XVII*[e] *siècles* (1913); the soaring of the Lyonese center was the object of a work by E. Vial and C. Cote, *Les Horlogers lyonnais de 1550 à 1650* (1927). The history of watchmaking in Geneva is now well known since the publication in 1916 of a study by A. Babel, entitled *Les métiers dans l'ancienne Genève. Histoire corporative de l'horlogerie (. . .)*. A better perspective on the importance of German horology in the sixteenth century is now available, thanks to recent studies by German historians such as K. Maurice and E. Groiss.

On the other hand, not one monograph has been published till now on the industry in Paris, despite the primordial importance of that center. This glaring omission stimulated our present work. The gathering of various documents gave us some insight into the life of Parisian horologists. The study of the regulations of the guild and the diverse edicts promulgated on its behalf, the examination of unpublished documents kept in the Archives Nationales, the reading of evidence, whether in manuscript form – such as that given by a case report found in the Bibliothèque Nationale or a report by Julien Le Roy on means for increasing the watch trade, which was discovered in the Conservatoire National des Arts et Métiers – or published in the *Journal des Savants*, the *Mercure de France*, the *Encyclopédie* of Diderot, the *Dictionnaire universel du Commerce (. . .)* of Savary des Bruslons or in watchmakers' writings, helped us discover the organization of the watchmaking trade in Paris, its evolution and its vicissitudes. Finally, the Brateau Archives in the Conservatoire National des Arts et Métiers were a precious assist.

Our research on the English industry was, without a doubt, more difficult. None of the very numerous English publications concerned with the history of watchmaking addresses itself to really detailing its historical development in that country. Fortunately, however, the study published in 1881 by S. Atkins and W. H. Overall, under the title *Some Account of the Worshipful Company of Clockmakers of the City of London*, was of help.

Our account of the watch's technical evolution does not claim to be a treatise or to describe in minute detail the multifarious inventions that benefited the watch, beginning at the end of the seventeenth century. We have merely tried to underline the stages in the conquest of precision and to examine the direction that, in accordance with their period, scientists and watchmakers took in their work. The study of reports relative to watchmaking and annually published in one volume from 1699 to 1790 in the *Histoire de l'Académie royale des Sciences avec les mémoires de mathématiques et de physique pour la même année (. . .)* was useful in helping us discern the fields of research that, at given times, attracted the interest of scientists and watchmakers.

Generally the work of technicians, very few studies on horological history reserve an important part to watch decoration. When that part exists, it is seldom seen from an art history perspective. Questions of primal importance, such as the search for initial models and the manner that watchcases reflect the style of a period, are neglected.

In order to approach the art of watch decoration, we resorted to the study of collections in French and foreign museums. For some of them, our knowledge was facilitated through the perusal of catalogues. In Paris, we examined the watch collections of, most notably, the Louvre, the Petit-Palais, the Musée Cognacq-Jay and the Cluny Museum (now kept in the Musée national de la Renaissance); in London, those of the British Museum and the Victoria and Albert Museum; in Amsterdam, those of the Rijksmuseum; in La Chaux-de-Fonds, that of the Musée International d'Horlogerie; in Rockford, that of the Time Museum. The assimilation of the analytical catalogue of the Louvre's Olivier Collection certainly marked a very important stage in the development of our research.

Our direct study of the objects was complemented by research, particularly in the Cabinet des Estampes of the Bibliothèque Nationale in Paris, on the models used by engravers and painters on enamel to decorate the watches: collections of engravings by ornamenters or etchings after fashionable paintings.

This history of the watch does not pretend, of course, to be exhaustive. On the contrary, it enabled us to perceive the extent of the work that needs to be done in the three domains it discusses. The development of many manufacturing centers – notably in England and the Netherlands – deserves more attention, so that their trading exchanges, production rates, work methods and guild organizations can be brought to light. Our knowledge of watchmaking's technical evolution could be made even richer through extensive research on the work of the great watchmakers, about whom, barring a few exceptions, we have all too brief and fragmentary information. In the domain of watch decoration, a good many questions remain obscure. The enamelists, goldsmiths and engravers whom the watchmakers would entrust with their work remain a crowd of obscure craftsmen of whose identity, places of work, rights and obligations we are ignorant. A historical and stylistic investigation of enameling appears necessary to us. We hope to have the possibility of soon presenting such a study.

Part one

Social and economic evolution

I From 1500 to 1600: The expansion of horology "in miniature"

The watch is not an invention, but the result of a reduction of the dimensions of the portable spring clock. The time and place that a portable clock was first built small enough to be carried on the person remains a mystery many scholars have tried in vain to solve. Indeed, it is essentially a spurious question. Clockmakers of several nationalities may have, simultaneously or at different times, come up with the idea of diminishing the volume of the table clock to make a clock small enough to carry on one's person.

In considering both the rare documents relating to small-work horology and the technical conditions necessary for building a watch, we shall see that one can reasonably establish the birth of the watch at the end of the fifteenth century.

All through the sixteenth century and until the end of the seventeenth century, the watch remained a luxury item, a preserve of kings, princes, grand bourgeois and the wealthy. The demand of this rich clientele stimulated the rise of watchmaking centers, some destined for a relatively short period of fame, such as Augsburg and Nuremberg, others for a brilliant future, such as Blois, Paris and Geneva.

The birth of the watch

The necessary technical conditions

What is required to make a clock that can be carried from one place to another without causing the stoppage of the movement? Two conditions are necessary: the development of a small driving mechanism whose running does not depend upon its position and the introduction of another part to regulate the driving mechanism, one whose action cannot be disturbed as the instrument is transported. Knowing the period when these conditions were met enables us to establish approximately when the watch first appeared.

As early as medieval times, armorers and locksmiths routinely made use of the elastic force of the spring. If one considers that the first clockmakers came from among these craftsmen, it is no wonder that some of them had, at least as early as the first years of the fifteenth century, the idea of also using springs in the making of clocks.[1]

A few documents bear witness to the existence of spring-driven clocks in the second quarter of the fifteenth century, notably at the court of Burgundy, which was renowned for its luxury.

A contemporary portrait of a Burgundian gentleman, Jean Lefèvre, lord of Saint-Rémy,[2] shows in the background a small clock shaped like a square turret, hanging from the wall by a chain. Through the open chassis, we can see a chime on the top and, at the base, the driving gears of both the chime and movement mechanisms, composed of a spring hidden behind the molding of the base and a fusee linked to it by a rope.

The same assemblage characterizes the wall clock that belonged to Philip the Good (1396–1467), duke of Burgundy. The clock, whose movement—if not the decoration—was made around 1430,[3] is now in the Germanisches Nationalmuseum in Nuremberg. An inventory of the duke's possessions, drawn on July 12, 1420, mentions a clock doubtless of a similar type: "Item: a small square clock, gilt on the outside, and its white enameled zodiac has one bell on top to ring the hours."[4] Dated 1423, the inventory of Margaret of Burgundy described "a small gilt clock. There are two panels on either side made of gilt silver, as is also the dial."[5] It is not unlikely that the small gilt silver clock of the duchess was of the spring-driven type, like the one that Temperance holds in a tapestry that portrays the Story of the Virgin (now in the Royal Palace in Madrid) by Jan van Eyck (1395–1440), a painter attached to Philip the Good.[6]

From the mid-fifteenth century onward, relatively numerous documents attest to the use of the spring-driven clock.

On July 18, 1459, Jehan de Lyebourg, "a master clockmaker living in Paris" received the sum of "96 livres 5 sols tournois" for five clocks delivered to King Charles VII (1403–1461), "four of which are with a bell and counterweights, and the other is but a half-clock gilt of pure gold and without counterweights."[7] A hexagonal clock owned by the Victoria and Albert Museum, on show at the British Museum, dates from this period. Its iron and brass movement has two springs contained in barrels.

On April 4, 1480, Louis XI (1423–1483) ordered payment to Jehan de Paris, clockmaker, "16 livres 10 deniers tournois . . . for a clock with a dial and that rings the hours, provided with all appurtenances, which the said lord sent for and had bought from him to carry about himself everywhere he goes." A year later, on June 5, 1481, Louis XI ordered paid to Marin Guerier "the sum of 50 solz tournois for having carried on his horse the clock of the said lord for ten entire days."[8]

Inventories taken in the second half of the fifteenth century give evidence of the growing use of table clocks. In an inventory of the furniture and jewels belonging to Margaret of Brittany, the first wife of Francis I, dated September 29, 1469, we see mention of "a small clock, garnished with gilt silver."[9] Among the possessions belonging in 1474 to the countess de Montpensier we noticed "a gilt silver clock in a red case" and "a clock in a leather case."[10] The

1 Illumination illustrating *L'Horloge de Sapience,* French translation of a work written in Latin by Arrigo Suso. Around 1450–1460.
The illuminator represented the diverse time-measuring instruments in use in the mid-fifteenth century: equinoctial dial, quarter-circle sundial, oval portable dial, shepherd clock, astrolabe, monumental weight clock, spring table clock. The representation of this movement, with its octagonal plates between which a fusee and a motive spring appear, is irrefutable evidence of the use of portable clocks, similar to those that have been preserved since around 1450 at least.
Brussels, Bibliothèque Royale, Manuscript IV, 111, fol. 13 vo.

luxury in the court of Charles the Bold (1433–1477), duke of Burgundy, is suggested by the inventory of his jewels. One is described as «ung aureloge d'or, assis sur six lyons, à plusieurs ymaiges à l'entour, garnye de XXXIX perles et XVII rubis, et au couppet a une salière sur laquelle a ung fusilz, pesant: VII m. VI O.»[11] ("a gold clock sitting on six lions, with several pictures around it, decorated with XXXIX pearls and XVII rubies, and inside it, a saltcellar container on which there is a fusee; weight: VII marks, VI ounces.") An illumination decorating a manuscript entitled *L'Horloge de Sapience* (The Clock of Wisdom), dated about 1460, shows a clock movement provided with a spring barrel and a fusee. Set between two octagonal plates, supported by posts, this movement is identical to that of the horizontal table clocks made in the sixteenth century and still extant.[12]

A letter written August 21, 1482, may serve as a commentary. A cleric practicing the art of clockmaking, Comino da Pontevico, was offering the marquis of Mantua one of his pieces, which he described like this: "The clock is provided with a steel spring that is hidden in a brass barrel, around which a gut cord is coiled. . . . [T]he gut cord is tied to the barrel attached to the steel spring in order to cause the traction of the screw [the fusee] to which it is tied in such a way that, thanks to the movement of that screw, which is produced by the steel spring, all the wheels of the clock are started. Such are my clocks. . . . [T]hus proceed the masters fabricating the clocks without weights, a few such models of which can be seen here in Mantua. . . . "[13]

In the same period, between 1475 and 1482, during a stay in Rome, Brother Paulus Almanus described in a manuscript thirty clocks, several of which used a spring drive and a fusee. One of them, indicating the hours and minutes, the date and the phases of the moon, belonged to the bishop of Mantua, Francesco Gonzaga. Was it perhaps one of the "models" made in Mantua that Comino da Pontevico was mentioning?"[14]

This evidence from diverse sources clearly attests to the existence of all the components necessary for the assemblage of watches in the second half of the fifteenth century. To obtain a clock suitable for carrying on one's person, all that was needed was the reduction of the dimensions of the works.

The dates of the watch's appearance

A letter from Jacopo Trotti, ambassador of the Este family to the court of Milan, written on July 19, 1488, to the duke of Ferrara, clearly mentions watches used to ornament three suits: "I want Your Lordship to know that M. Lodovico is secretly having made three silk suits richly enhanced by very pretty pearls. All are of the same cut and are ornamented with a watch with little bells, except the one of M. Lodovico, who does not want his to strike, for he wants to be the cause for the others' ringing. And to each festive suit are affixed two short lines of verse, which you will find herewith on this little sheet. . . . " Here are these lines translated from the Spanish:

For M. Lodovico
Though the watch here does not ring
Satisfaction the work does bring.

For the duke
The delicacy of this labor
Makes the watch ring out the hour.

For M. Galeazzo
As long as this watch runs well
My repute will also swell.[15]

Of fundamental importance, this document, brought to light by Enrico Morpurgo, confirms what the preceding analysis tended to suggest.

Among the many ivory objects owned by the countess of Montpensier, inventoried in 1474, we note "*une monstre d'oreloge d'yvoire*" (an ivory clock-watch). It is not unlikely that this designates a watch having a case made of ivory. We find this wording (*monstre d'oreloge*) applied to watches in several legal documents of the sixteenth century. Furthermore, the ambiguity of the expression, which might be applied as well to a sundial, is erased by the fact that the inventory also mentions that kind of object ("three sundials, two of silver and one of ivory, and another ivory sundial"). As we have pointed out elsewhere, Gabrielle de La Tour—who must have had a certain taste for timepieces—owned two table clocks designated by the word *oreloge*.

These different pieces were kept, at the time of the inventory, in boxes and chests. The "*monstre d'oreloge d'yvoire*" was enclosed in a "flat, square case" that contained five other personal objects, among them a small mirror and a fork.[16]

The growing use of the watch

From the year 1500 onward, the use of the watch spread in the courts of Western Europe. One of the first users was King Francis I (1494–1547). On December 31, 1518, he ordered: "Jean Sapin, tax collector general of Languedoc and Guyenne, shall pay Julien Couldray, horologist of Blois, the sum of 200 écus d'or au soleil for his payment of two excellent daggers adorned on the pommels with two entirely gilded *horloges*, reserved for the King's usage."[17]

Other orders executed by Julien Couldray for the king—who had taken him into his service as a valet de chambre and horologist—are known. In 1523 he delivered to him "two clock-watches without counterweights for the service of his room."[18] On March 26, 1530, he received fifty écus to start a clock-watch.[19]

The entourage of the king of France also showed interest for this novel timepiece. Made in 1532, the objets d'arts inventory of his minister, Florimond Robertet, included "twelve watches, seven of which ring, and the five others are silent with cases in gold, silver and brass of various sizes."[20]

Unpublished documents kept in the Archives Nationales in Paris confirm the relatively routine production of watches in the second quarter of the sixteenth century.

An estate inventory, providing us with a description of the shop of a Parisian watchmaker, Nicolas Morel, records the presence, on December 9, 1534, of a "clock-watch in gilt brass, furnished with a case also gilt, locking with a key," "two started small, round clock-watches," "a dozen and a half undented wheels for use in clock-watches," and "two small gilt watchcases."[21] Another death inventory, dated January 27, 1547, enumerates in particular: "Item: two small gilt watches. . . . Item: another small gilt watch, furnished with its own iron case. . . . Item: one movement of a round clock-watch, furnished with a case of gilt leather."[22]

One of the first watchmakers whose name has been preserved is a locksmith of Nuremberg, Peter Henlein (c. 1479–1542), a master since 1509.[23] His talent is praised in an edition of the *Cosmographia Pomponii Melae* published in 1512 in Nuremberg by Johann Cochlaeus, schoolmaster and theologian, in these words: "*Inveniuntur in dies subtiliora: etenim Petrus Hele, juvenis adhuc admodum, opera efficit, quae vel doctissimi admirantur mathematici, nam ex ferro parvo fabricat horologia plurimis digesta rotulis, quae, quocumque vertantur, absque ullo pondere et monstrant et pulsant XL horas, etiamsi in sinu marsupiove contineantur.*" (Every day, they invent more subtle things: It is so with Peter Hele, still almost a child, who does work that even the most learned among the mathematicians admire, for, with a little iron, he produces clocks made up of many wheels that, in any position and without any weight, show and strike forty hours even when worn on the breast or in the pocket.)

A cylindrical watch, kept in Nuremberg's Nationalmuseum, bearing the inscription "Petrus Hele me f. Norimb. 1510," is without a doubt similar to the portable clocks described by J. Cochlaeus, but its attribution to Peter Henlein must be rejected—as Klaus Maurice has so keenly pointed out—being founded on an inscription added at a later time, its characters directly copied from the printed text of the *Cosmographia Pomponii Melae*. An archival document, on the other hand, preserves the memory of a golden musk apple containing a clock movement, for which fifteen florins were paid to Peter Henlein in January 1524.[24]

Margaret of Austria (1480–1530), from 1507 the regent of the Netherlands, owned a watch of that shape. In July 1523 the inventory of jewels, objets d'art, paintings and books she kept in her palace of Malines mentions a "small clock shaped like an apple, with a small gold chain, has a ring at the end of the chain."[25]

As these few documents reveal, the first watch owners were princes and great lords.

A princely clientele

While belltower clocks responded to a collective demand in which public interest and civic pride were generally combined, the making of portable clocks satisfied an individual need.

The search for utility was obviously one determining motive: the campaigning prince certainly appreciated keeping time for the better ordering of his strategy; the traveler liked the advantage of owning a watch, not needing to wait until he'd crossed a town with a clocktower to know the time. The dilettante or scholar also liked to have on his worktable a timepiece to organize his schedule.

The taste of the Renaissance princes for mechanical curiosities also contributed to the development of watchmaking. In his letter to the marquis of Mantua in August 1482, Comino mentions that the spring drive of portable clocks is hidden inside a barrel and

2 Hans Holbein (Augsburg, 1497–1498—London, 1543), *Portrait of Georg Gisze,* 1532.
The subject, a rich London merchant, is shown at his worktable, on which there is a small table clock through whose open shutter the winding mechanism is seen. Its dimensions (about five centimeters in diameter, four centimeters in height), which can be estimated, thanks to the ring nearby, and its shape are identical to still extant pieces of French origin. We may note that in this portrait the clock, accented with flowers, has the symbolic value of a memento mori.
Berlin-Dahlem, Gemäldegalerie.

emphasizes that "all the masters build this instrument so as to make it more beautiful and mysterious".[26]

This taste favored the creation of watches surprising for their complications or the difficulty of their execution. For instance, the prince of Urbino received a ring crowned with a gem in which a watch movement was set. Cardanus gives this description: "A ring to put on the finger near the thumb, on which there was a precious stone that had a complete clock that, in addition to the line that distinguished the hours, struck for the one who wore it, at hourly intervals."[27]

To the quest for the useful and the mechanically interesting, was added the pride of owning a rare object. Many princes chose to be portrayed with a table clock or a watch. Thus, a small table clock often appears among the familiar objects surrounding the personages painted by Titian (c. 1490–1576) and Hans Holbein (1497–1543).[28] The portrait of Georg Gisze, painted in London in 1532, kept in the Gemäldegalerie in Berlin–Dahlem, shows a small cylindrical clock set on the model's worktable. This piece is similar, in shape and dimensions, to several pieces still in existence, such as the small cylindrical table clock executed during the same period by the Parisian watchmaker Antoine Beauvais, owned by the Time Museum, or the one signed *Noël*, dated 1504, which belongs to the Olivier Collection in the Louvre (inv. OA 8397).

A few portraits done around the mid-sixteenth century, recorded by F.A.B. Ward, include the display of a watch.[29] One, done by Jacopo da Pontormo (1494–1556?), dated 1549, presents the picture of a drum-watch with a pierced lid, such as we shall describe in the section devoted to the evolution of watch decoration (private collection). Another, done in 1550, attributed to Ludger Tom Ring the Younger (1522?–1584), shows a young man holding a spherical watch characteristic of watches produced toward the mid-sixteenth century (private collection).

The introduction of watches into the daily life of the privileged few was also promoted by the taste for luxury. During the sixteenth century, watches became articles of finery, thanks to techniques borrowed from the arts of the goldsmith, jeweler, and enameler that were applied to their decoration.

Queen Elizabeth I of England (1533–1603) was particularly responsive to the subtleties of watch decoration. The inventory of her jewels, drawn up in July 1587, indicates that she owned a large number of them, àll lavishly decorated with diamonds, pearls, gems and colored stones.[30] Let us admire: " . . . One litle watch of golde gar. [garnished] on the border with very small sparkes of rubies and emerodes wth christall on both sides and a pearle pendant gar. with golde like a flehe flye"; "a litle watch of christall slightlie gar. wth golde wth her Ma^ts [Majesty's] picture in it"; "a watch of Agatt made like an egg garnished wth golde"; "a litle watch of golde thone side wth a frogge in the topp, thother side gar. with small garnettes like a pomegranett."

The first watchmaking centers

Under the influence of growing demand, ever greater numbers of craftsmen specialized in the art of watchmaking. During the sixteenth century, certain towns in Germany and France, favored by the proximity of a court or benefiting from an important commercial activity, became genuine watchmaking centers, thanks to the relatively high number of watchmakers grouped within them.

Italy

The cities in Northern Italy counted skillful horologists during the medieval period and in the sixteenth century. Unfortunately, their names are associated only with brief contemporary allusions or archival documents.[31] Not one clock survives of Cherubino Sforzani (c. 1485–c. 1558), of Gianello Torriano of Cremona (c. 1515–c. 1585), the clockmaker of Emperor Charles V,[32] of Lorenzo della Volpaia of Florence, the creator of a planetarium which once was in the Palazzo della Signoria; of Giovanni Giorgio Capobianco (c. 1490–c. 1555), a clockmaker renowned for his automata and his watches. In that galaxy of ingenious mechanicians, as capable of building clocks and automata as of organizing public works, Pietro Guido (c. 1450–c. 1510), a native of Mantua, seems, in contrast, to have specialized in horology "in miniature". According to Morpurgo,[33] he was even nicknamed "dell'Orologio" because of his specialization.

Despite their activity, which was no doubt considerable, the Italian watchmaking shops never succeeded in forming real production centers comparable to those that appeared, in the course of the sixteenth century, in Germany and in France.

Germany: Nuremberg and Augsburg

During the Middle Ages, the craftsmen in Nuremberg and Augsburg had gained a strong reputation in the metal arts as goldsmiths, armorers and locksmiths. They exported their products to numerous foreign countries. It is thus not surprising that as early as the start of the sixteenth century, several among them, including the locksmith Peter Henlein, became interested in the technique of medium horology and horology "in miniature", which could open larger markets to them than could the big belltower clocks.

The two cities were horological centers of the first importance during the sixteenth century and the beginning of the next. Most of the watches shaped like drums and dating from the sixteenth century that have come down to us are of German origin. The relative abundance of these watches, in relation to the number of sixteenth-century Italian or French pieces that remain, attest in a convincing manner to the importance of the production of these two locations during that period in the domain of small-work horology.[34]

This art was freely practiced in Nuremberg until 1565, when a decree spelled out what masterpieces were required from horologists "in miniature" to obtain their masterships. The candidates were obliged to present a table clock striking each hour and the quarter hours, with calendar and astrolabe, and also an alarm watch.[35]

In Nuremberg, as in other German towns, the master clockmakers belonged to a guild that also counted as its members the armorers and locksmiths. There is unfortunately no record of their number during the sixteenth century.

Thanks to a study by Eva Groiss,[36] based on unpublished archival documents, the horological industry that developed in Augsburg during the sixteenth and seventeenth centuries is better known. The clockmakers in this city belonged to the Blacksmiths' Guild, which included also the locksmiths, armorers and bellmakers. A strict system of regulations was established between 1552 and 1554, covering their obligations and their rights. The usual corporative rules figured here. The apprentice had to be at least twelve years old to be hired and taught the trade for three years. After his apprenticeship, he had to become a journeyman for six or seven years. Only then could he attempt to become a master by producing a masterpiece. Six months were allotted to him to devote to this work. During the period, he could neither spend any of his time on other work nor get married. Conversely, in order to settle down as a master and establish a shop, a watchmaker was obliged to get married and to belong to the Blacksmiths' Guild (*Schmiedgerechtigkeit*). The sons of masters, as in many other places, received many privileges.

The type of masterpiece the journeyman was expected to execute was detailed in 1558; then, for a second time, in 1577. This regulation was observed until 1732. In 1558 the candidate was required to make a table clock, about twenty-three centimeters high, without any weights, striking each quarter and including an astrolabe. In addition, he was asked to produce a small horizontal or spherical clock that indicated the phases of the moon. In 1577 a list was drawn up allowing the journeyman to choose from among five kinds of clocks, all incorporating the same complications but differing in their form. In any case, the masterpiece was supposed to have a striking mechanism for the hours and quarters, an alarm, an astrolabe, the indication of the month and date, the planets and their signs. In outward form it could have the appearance of a clock about twenty-three centimeters high, a "monstrance" clock (*Spiegel*), a square-shaped clock, a hexagonal clock, or a clock in the shape of a square tower, topped by the astrolabe. As in 1558, the making of a small horizontal or spherical clock was also required.

From 1500 to 1600 the number of watchmakers in Augsburg increased constantly. Between 1500 and 1550, sixteen watchmakers belonged to the Blacksmiths' Guild and thus were able to establish a shop. Between 1550 and 1600, this number rose to eighty-one. In 1582 the city had forty-six watchmakers and fifteen journeymen. Such an increase was tied to a mounting demand for clocks and watches made by the masters in the city, whose talent is evidenced by the pieces that have survived.[37]

The court of Constantinople had a great appreciation for their products, which the emperor sent there with great regularity as gifts. In October 1573 the imperial envoy announced the grand vizier's wish to own a few of these watches.[38] Two months later, Maximilian II ordered made three watches to be worn around the neck. The watches, whose shapes imitated the musk apples, particularly prized in the East, met with instant success in the sultan's court. A precious attestation of the fact is given by two extant drawings,[39] dated 1576, showing two models of the spherical watches ordered by the grand vizier. The first model, decorated with foliage and having a dial with Turkish numerals, was to be made in four reproductions. The second, decorated with interlacings, had Roman numerals and was to provide three reproductions, two with alarms.

The German Empire numbered other cities in which the watchmaking industry was quite active: Vienna, Dresden, Innsbruck, Munich and Strasbourg.[40]

France: Paris, Blois and Lyons

In France the expansion of horology "in miniature" was brought about by the demand of a clientele composed of princes and high lords. The presence of the court in Paris, Blois and Lyons caused workshops to spring up, becoming numerous enough to form real watchmaking centers.

In Paris the industry reached such proportions that a guild was established in 1544 at the request of six watchmakers: Fleurent Valleran, Jehan de Presles, Jehan Pantin, Michel Potier, Antoine Beauvais, Nicolas Morel, Nicolas le Contançois.[41] Besides these masters, many other watchmakers were active in Paris in the second quarter of the sixteenth century. Archival documents show the names of Mathurin Benoist, Michel Bertrand, Martin Bezelin, Jehan Bourillon, Jacques Fieffé, Loys François, Jean de Laulne,

Jean Lignyer and Blanchet Morant. Some documents give us an outline of the careers of these masters.

Jehan de Presles is designated the king's "*horologist*" in a document dated August 29, 1546.[42] As early as 1539, he was in charge of the maintenance of the palace clock; he kept that position at least until 1556.[43] On December 9, 1534, he was an expert for the death inventory of Spire de Fons, wife of the watchmaker Nicolas Morel.[44]

In 1541 Michel (or Michaud) Bertrand was also named the king's watchmaker.[45] Other documents dated in 1534, 1539 and 1546 mention him.[46]

Mathurin Benoist also had the title of king's horologist. On May 6, 1547, he hired Jacques Gremy, another watchmaker, as his journeyman for four years.[47]

In 1528 Loys François worked as both a locksmith and watchmaker. This fact is established by an annuity agreement signed with his wife, Guillemette Florentaine,[48] on May 2 of that year. Six years later, in his capacity of master horologist, he was another expert in the death inventory of Nicolas Morel's spouse. In 1538 he was again designated to the same function after the death inventory of his colleague's second wife.[49] We think it is possible to attribute to him a French clock in the shape of a hexagonal tower, now in the British Museum, which has the signature "Lois F" on the bottom and an estimated manufacture date of between 1530 and 1540.[50]

Nicolas Morel is known through several archival documents. The oldest one, dated December 9, 1534, is the inventory taken after the death of his wife, Spire de Fons. This document is all the more interesting in that it details the conditions of the master's shop on the rue des Arcis. Before four years had gone by, Nicolas Morel was assisting at the death inventory of his second wife, Martine de Moutonvilliers, on October 14, 1538. His home address was "rue des Juifs."[51] This clockmaker seems to have reached rather easy circumstances, if one can judge from his purchase of two vineyards, one in Sceaux, the other in Clichy.[52]

Two master watchmakers were present at the death inventory of Barbe Nepveu, the wife of Master Hilaire Josse, surnamed de la Chapelle, barrister of the Parlement of Paris: Jacques Fieffé and Jehan Bourillon.[53] The latter is known by another legal act: on January 15, 1548, he hired as an apprentice Jean Greban, who also became a master.[54]

Blanchet Moran was both a master watchmaker and a maker of harquebuses. He lived with his wife, Mondette Loupiac and his two daughters, Martine and Claude, on "rue Sans-Chef, off the rue Saint Antoine." These facts are given in the inventory of his wife's estate, dated March 30, 1545.[55]

Martin Bezelin, who lived on "rue de la Vieille Pelleterie," died in 1547. His inventory was drawn up on September 24 at the request of his widow, Simone Alorge.[56]

A notarized act pertaining to the auction of the contents of a watchmaker's shop in the Grand Courtyard of the Palais de Justice, dated October 24, 1547, preserves the name of Jean Lignyer, watchmaker.[57]

In 1549 Fleurent Valleran and Nicolas le Contançois were the visitor-wardens in charge[58] (see below). The same year, on June 16, ten master watchmakers were part of the cortege during the royal entry of Henry II (1519–1559).[59]

Some very rare pieces that have survived attest to the work of the Parisian masters. Two have already been mentioned: the table clocks signed by Antoine Beauvais and Loys François. Two others, signed by Fleurent Valleran, belong to the Musée International d'Horlogerie (inv. no. 1) and to the Wallace Collection (inv. III J 532). A third, signed by Nicolas Morel, is in a private collection.

The horologists requesting the formation of a guild stressed that it was ". . . necessary for the public good in our City of Paris, capital of our Kingdom, that there be expert persons, well apprised and knowing securely the work and labor or art and craft of the horologist, and that they do these works with good materials and stuffs, to obviate the abuses, malfeasance, faults and negligences that daily were and are done and committed by several of the said horologist's trade."[60]

Fourteen articles regulated the new guild, whose statutes were promulgated by Francis I in July 1544. The apprenticeship was fixed at a minimum six-year duration. No one could practice the profession without presenting a successful masterpiece: "They who would presently be masters of the said craft shall be required to make a masterpiece of the said craft, such as shall be ordered by some of the elder masters and the most experienced or others

3 Table clock. Engraved gilt brass case, signed "Beauvais." Paris, c. 1540–1550. D. 51.
This small spring-driven cylindric clock was made by a Parisian master, Antoine Beauvais, active in the second half of the sixteenth century. In its dimensions and shape, it is similar to two table clocks now in the Louvre (inv. OA 8282 and 8397) and to the one shown in the *Portrait of Georg Gisze*.
Rockford (Illinois), Time Museum, inv. no. 1070

4 Table clock. Engraved gilt brass case, signed by Fleurent Valeran. Paris, c. 1540–1550. D. 66; H. 58.
The miniaturization of the table clocks like this one and the one of Beauvais gave birth to drum-shaped watches.
La Chaux-de-Fonds, Musée International d'Horlogerie, inv. no 1

5 Round watch. Gilt brass case, dated 1548 and marked with the initials *C. W.* D. 56; H. 25.
The watch was probably made by a Nuremberg master, Caspar Werner, active from 1527 to 1557. Its iron movement has a stackfreed and a straight foliot. It is the earliest-dated watch known.
Wuppertal, Uhrenmuseum

3

4

5

6 Oval watch. Engraved silver and brass case. Movement signed "Charles Peiras A Bloys." About 1600. L. 46; W. 36.
The inside of the silver lid represents Ceres crowned by a winged genie. The hour circle is applied on a brass plate, Saint Anthony kneeling among his swine engraved in the center.
Paris, Louvre, Garnier Coll., inv. OA 7020

keeping presently an open shop of the said craft in our City of Paris. . . . And if they are found sufficient and so reported by them, they shall be received as masters of the said craft." The article that followed dictated: "Let no one of any condition whatsoever, if they are not received as it is said, be permitted to make or have made clocks or alarm clocks, large or small watches, or any other work of the said horologist's craft in the town, city and suburb of the said Paris, under pain of confiscation of the said work and an arbitrary fine applicable as stated." Only the sons of sworn masters could be received as masters without having made a masterpiece.

Masters were obligated to work in a shop open to the street: "[It is ordered] that the said masters shall not be able to work at the said craft if they do not keep a shop . . . opening on a public street." They had the obligation of marking their work: "The said sworn masters of the said craft are bound to choose a mark, which they shall declare to the said Wardens and Visitors. . . . " Merchandise coming from another town in the kingdom or from abroad to be sold in Paris was to be inspected by the "visitor-wardens": "[It is ordered] that any traveling merchandise of the said craft that shall be brought and carried from any place whatsoever, from inside or outside our Kingdom, to our said City of Paris, to be there sold either in bulk or by the piece, shall be visited by the said Wardens and Visitors under pain of confiscation of the said merchandise so prohibited and of arbitrary fine. . . . " Horologists' widows could keep on managing their husbands' shops, as long as they were assisted by able men "expert at the same craft."

The swift expansion of the industry in Blois started with the works Julien Couldray executed for Louis XII and then for Francis I.[61] Their high quality earned Couldray the privilege to wear the title of king's horologist. Upon his death around 1530, two other Blesians were also granted that office: Jean du Jardin and Guillaume Coudray. From that time on, the watchmakers' shops multiplied in Blois. Abbé Develle makes this statement: "If under the reign of Francis I three or four shops were started, seven new ones appeared under Henry II and eighteen at least under the reign of Henry III; and still we mention only those whose existence we have proven indisputably."[62]

From the sixteenth century on, the growing renown of the new watchmaking city attracted men of the craft from diverse towns in the kingdom, such as Abel Bérault, a La Rochelle native, who arrived in Blois in 1593, and Thomas Ribard, born in Rouen, who came around 1600.[63] Foreign craftsmen also came to settle in Blois during the sixteenth century; among these, let us mention Paul Cuper and Marc Girard, both arriving from Germany, and Jacob Deburger, originally from England.

At the end of the century, Blois numbered about twenty watchmakers: Abel Bérault, Salomon Chesnon, Robert Chevallier, Paul Cuper the Elder, Paul Cuper the Younger, Pierre Cuper, Jacob

Deburger, Jacques Enguérant, Marc Girard, Simon Gribelin, Abraham de La Garde, Jean de La Garde, Nicolas Lemaindre, Louis Leveulle, Charles Peiras, Pasquier Peiras and Louis Vaultier. Some of these names are not only mentioned in archival documents found by Develle, but are also inscribed on the plates of a few watches that have reached us, fancy watches of the most diverse shapes, oval or octagonal watches with their cases delicately engraved with allegories, mythological or religious scenes in the style of Etienne Delaulne.[64] Charles Peiras, Pasquier Peiras, Paul Cuper and Louis Vaultier are represented by such watches in the Paul Garnier Collection in the Louvre. Nicolas Lemaindre is represented in the British Museum by an oval watch with a plain back; Louis Vaultier, by a watch with a pierced and partly enameled back.[65] Marc Girard is represented in La Chaux-de-Fonds by an oval watch that also gives astronomical indications.[66]

The increasing number of watchmakers, the will to protect the trade from malpractice by unscrupulous or unqualified adventurers, and the relentless competition from some other guilds, such as the goldsmiths', are the reasons that justified the creation in 1597 of a Blesian Horologists' Community. Its statutes were almost identical to those of the Parisian guild.[67]

From the mid-seventeenth century on, small-work horology developed well in Lyons. The prosperity of the city, made rich by its trade, and the frequent presence of the court had combined to create a strong demand for luxury items. According to Vial and Cote,[68] six or seven horologists were active in Lyons between 1550 and 1560. Ten years later, there were thirteen. From 1580 to 1600, there were ten.

7 *Left.* Oval watch. View of the movement signed "N Lemaindre A Bloys." First half of the sixteenth century. L. 49; W. 36.
Nicolas Lemaindre, watchmaker and valet to Queen Marie de Médicis, was one of the most renowned Blesian masters. Watches and archival documents preserve his memory.
Right. Oval watch. View of the movement signed "Noël Cusin A Autun." First half of the seventeenth century. L. 48; W. 41.5.
Noël Cusin, born in 1587, succeeded his father, also named Noël, as evidenced by a May 1625 notarized act. He was still living in 1656.
London, British Museum, inv. 88, 12-1, 183; 74, 7-18, 18

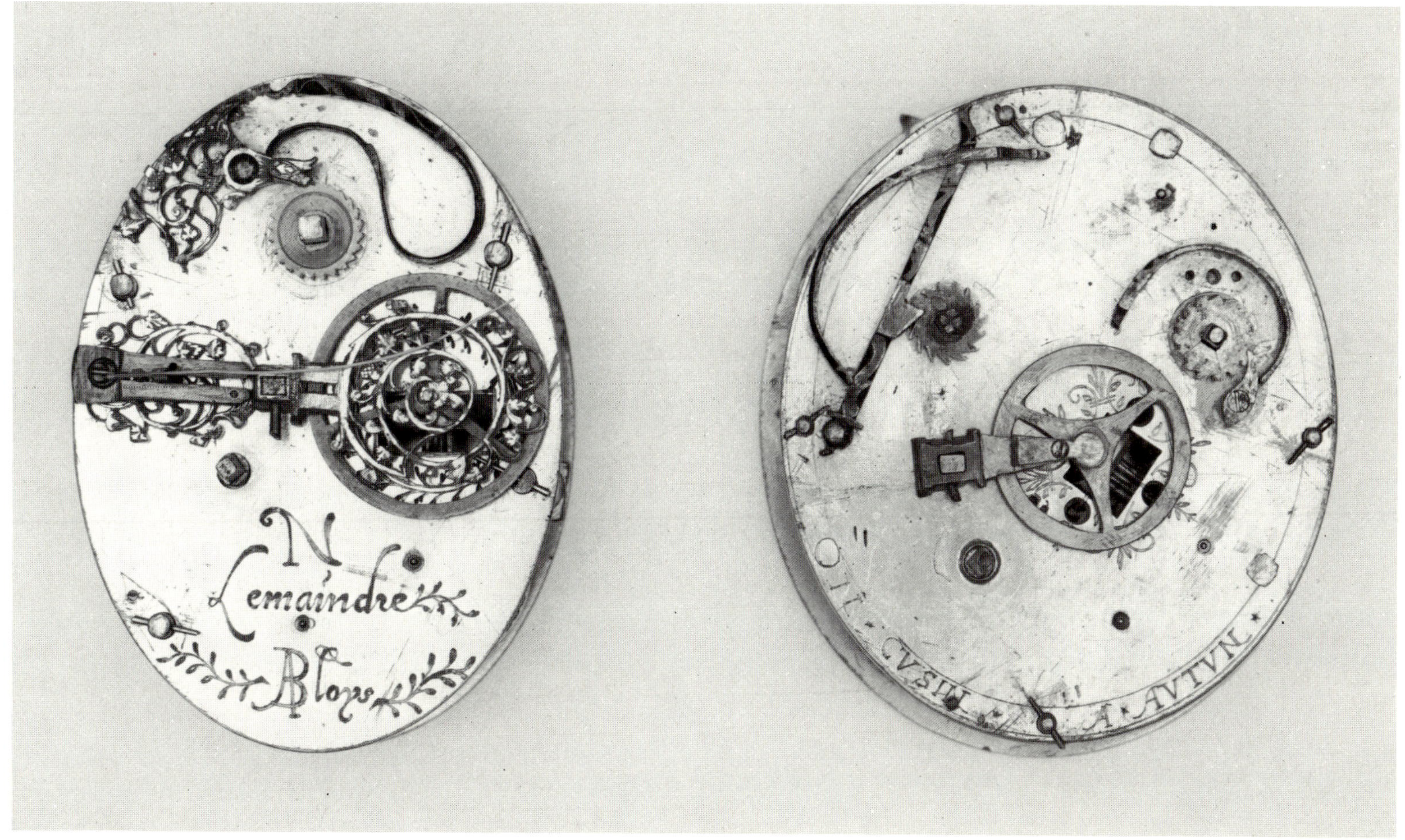

8

9

10

One of the first watchmakers to establish a shop in Lyons (in 1554) was Jean Naze, a native of Picardy. He died there of the plague in 1581, after leaving his possessions to the poor at the hospital and the public alms-house. The clocks and watches that were part of the donation were distributed during a lottery in 1592.[69] The inventory mentions six watches worth from eighteen to seventy-five livres, five of which were "made in ovals," and several clocks with chimes and astronomical indications.[70]

Most of the Lyonese horologists whose memory is preserved in archival documents or through a piece of work in a collection were active from the last quarter of the sixteenth century to the early seventeenth century. Among these, let us cite, for instance, Christophe Noytolon, born in 1558, who established himself in the trade in 1586, and Pierre Combret, who practiced his trade from about 1580 to his death in 1622. Several watches, carrying the signature of Pierre Combret have been preserved: an oval watch in the Louvre, an octagonal watch in the Musée de Dijon, a rock crystal watch in the Petit-Palais, and a cruciform watch in the Metropolitan Museum.[71]

All through the sixteenth century, the art of small-work horology developed in several French cities and towns. Thus, in Autun the Cusins managed a well-known shop that produced numerous watches:[72] one of oval shape, with a plain watchcase, can be seen in the British Museum (inv. 74–718–18). The use of watches in Autun doubtless began relatively early: Anatole de Charmasse, indeed noticed in the diary written by Canon Gaucher a passage, dated January 13, 1540, mentioning "*une monstre*."[73] The horologist Guy Rumault practiced his craft in Abbeville; the Musée International d'Horlogerie owns a watch bearing his name (inv. 500).

Watchmaking had flourished so well in sixteenth century France that from that time on, and particularly during the Protestant persecutions, it provided foreign countries with large contingents of watchmakers, who established their workshops in cities until then deprived of that industry. Thus, Geneva benefited from the migration of several French watchmakers as early as the second half of the sixteenth century.

8 *Left.* Oval astronomical watch. Engraved silver and brass. Movement signed "Marc Girard A Blois." Early seventeenth century. L. 55; W. 45.
Made of gilt brass, engraved with foliage scrollwork and two birds and two rabbits, the dial offers, besides the time, the following indications: the days of the week with their symbolic figures, the phases and age of the moon, the date, the months with the numbers of their days. The movement has an *au passage* striking train.
A native of Germany, Marc Girard married Anne Papin, a joiner's daughter, in 1593, the very year of his arrival in Blois. He died between 1615 and 1617.
Right. Oval watch. Engraved silver and brass. Movement signed by Guy Rumault. Abbeville, early seventeenth century. L. 55; W. 48.
The brass dial is engraved with scrollwork peopled in the style of Delaulne and Jacquard. The movement has an *au passage* striking train.
La-Chaux-de-Fonds, Musée International d'Horlogerie, inv. 556 and 500

9 Cruciform watch. Cut rock-crystal case. Movement signed "P Combret A Lyon." About 1600. L. 56; W. 35.
Pierre Combret, who died in 1622, is one of the rare Lyonese watchmakers to have his name kept alive by several watches.
Paris, Musée du Petit-Palais, Dutuit Coll., inv. 1411

10 Peter Claesz (1598–1661), *Still Life with Musical Instruments,* 1625.
Dutch painters often inserted a watch in their still lifes as a reminder of the flight of time. The one presented by Claesz, a Haarlem artist, is typical of watches produced in his time.
Paris, Louvre.

The beginnings of Genevan horology

The beginnings of Genevan horology, closely tied to the establishment of the Reformation, were fueled by diverse social and economic conditions.[74] A little after 1550, Geneva opened wide its doors to all who were fleeing religious persecution. Foreigners were welcome, for the city, after going through a very prosperous period, of which its famous fairs were proof, had been losing a large part of its population. The decline had started with the protectionist measures taken by the French king and with the growth of important trade centers, such as neighboring Lyons. The welcome given the refugees was the more friendly in that they were, by and large, qualified craftsmen. The sudden influx of religious exiles had fortunate consequences on the economic life of the city, which soon regained its commercial power. Traditional industries, such as hosiery, clothmaking and jewelry, went soaring, while new industries, such as silk and velvet manufactures, lacemaking and watchmaking, were growing strong roots.

Among these activities, watchmaking was to undergo an exceptional expansion. A. Babel once wondered why the art had attained such a supreme place in the life of the city. To start with, he saw in the already advanced development of the goldsmith trade a condition propitious to the brilliant expansion of the watch industry. The many goldsmiths in the city indeed made its settling there favorable: the new industry would have to call on their talents, since they would be keeping for themselves the arts of manufacturing and decorating watchcases, which were at that time real jewelry. Some of them had already added watchmaking to their own craft when the Genevan Horologists' Guild was created in 1601. Elsewhere, Babel very aptly insists on the fact that industries like watchmaking and jewelry, demanding a large amount of labor and very little material, could easily develop in a city like Geneva, which lacked raw material but was densely populated: "Raw materials that are easy to transport and that demands a large quantity of labor: that is what Geneva needed."[75]

The persecutions that followed the Edict of Chateaubriant (1551) and later that of Compiègne, marked the start of the exile of French horologists. François Somellier, a native of Dieppe, arrived in 1555 and was accepted as a burgher on January 31, 1559. Pierre Lodereau, originally from Nérac, also took refuge in Geneva in 1557. Clément Bergier, born in Lyons, received his burghership on the same day as Somellier.

After the bloody night of the Saint Bartholomew massacre of the Protestants (August 24–25, 1572), other watchmakers came to settle in Geneva. The name of Charles Cusin, accepted as an *habitant* on July 12, 1574, is famous in Geneva, for local tradition attributes to him the introduction of watchmaking in the city.[76] His father was Noël Cusin, a master horologist who had managed an important shop in Autun. Charles's skill attracted the attention of the Council of Geneva, which ordered him to invent a system to improve the ringing of the bells of Saint-Pierre. His work being judged satisfactory, he was received without the usual fee as a burgher in Geneva, on May 15, 1587. During the last quarter of the century, Geneva continued to welcome the French watchmakers, among them Le Melais, a native of Paris, and Jacques Fremin, the Parisian locksmith who built the clock on the bridge over the Rhone.[77]

The exile of the French craftsmen as a consequence of the religious persecutions during the second half of the sixteenth century appears to foreshadow the decline of French horology under the reign of Louis XIV, after the revocation of the Edict of Nantes, which would also provoke the departure abroad of a very large number of Huguenot clock- and watchmakers. Nevertheless, for the time being, the birth of important horological centers like Paris, Blois and Lyons and the development of the art in various cities and towns in France bore witness to the vitality of French watchmaking and set the stage for the exceptional flowering it would undergo during the first half of the seventeenth century.

II From 1600 to 1660: The strengthening of the watchmaking industry in France

During the first two-thirds of the seventeenth century, the map of the centers that would impose their watchmaking production on the world started to take shape. In that period, French horology went through a period of unrivaled development: the art acquired its noble status in Paris, while it was gloriously blooming in Blois. Meanwhile, two watchmaking centers, Geneva and London, were growing, preparing their bid for supremacy.

The prestige of French watchmaking

In the first half of the seventeenth century, French watchmaking went through a remarkable period of development due to the support of the royal government and the widening of its clientele. To the local demand of the nobles and rich bourgeoisie was added the strong demand coming from abroad. The horological centers that had appeared during the sixteenth century, Paris, Blois and Lyons devoted much of their production to watches and small clocks. Moreover, the art was also prospering in numerous other towns, such as Rouen and La Rochelle.

Paris

The Parisian horologists felt the need to demand new statutes assuring a better protection of their rights against the other guilds and a more stringent selection of masters of the craft. A twenty-four-article list, signed by the master craftsmen Beton, Margotin, Bernard, Gamot, Beauvais and Raillart, was presented in March 1645 to the king's counselor, Dreux Daubray, and to the royal prosecutor of the Châtelet, Etienne Bonneau. The articles, being judged "just and reasonable," were then approved on February 20, 1646, by the young Louis XIV.[78]

The master horologists of Paris thus obtained official confirmation of their privileges. In the first place, numerous favors were granted to those of their sons who wished to undertake the same career as their fathers. These measures restricted the number of candidates coming from other professions and candidates in more modest circumstances. Article 7 ordered: "The sons of masters who are able will be preferable to apprentices: on pain of a two-hundred-livre fine for the wardens of the said trade who will contravene," and Article 19 added: "No one shall be admitted to the mastership, unless he pays to the purse the sum of forty pounds . . . save the masters' sons, who will not pay anything."

The number of master horologists was limited to seventy-two in the city of Paris (Article 7). Attesting to the development of the watch trade in Paris during the first decades of the century, this decree aimed to close access to the guild. Article 8 reinforced that restrictive measure: "Let no one be admitted or receive his mastership in the said horological art unless he has done his apprenticeship with a master of our said city of Paris."

The duration of the apprenticeship was fixed at eight years (Article 3). The master could not train more than one apprentice at a time; he had, however, the right to take a new apprentice when the first had finished his first seven years of apprenticeship. This measure was meant not only to better the training of future masters, but also to limit the number of candidates for the mastership.

No one could be accepted as a master watchmaker in Paris if he had not made a masterpiece, which had to be, at least, a clock *"à réveil-matin"* (Article 6) that is, an alarm clock.

Several articles affirmed the rights of the watchmakers over the claims of the goldsmiths, who tirelessly had been struggling against them, to prevent watchmakers from working with precious metals. The struggle was harder for the watchmakers, who were less influential and much less numerous than the goldsmiths.

Article 13 forbade goldsmiths from manufacturing horological objects or from trading in them: "[It is ordered] that it shall not be permitted to any goldsmith or anyone of any estate and trade whatsoever to engage in working or trading, directly or indirectly with any merchandise of horology big or small, old or new, finished or unfinished, if he has not been received master in the said art of Horologist in our city of Paris. . . . "

The horologists were free to make or to have made in any kind of material their watch or clock cases: "[It is ordered] that the master Horologists shall be able to make or to have made all their said works of horology, as much the cases as any other parts of their said art, of such stuff and material that they can think good for the embellishment of their said work, as much of gold as of silver and other stuff that they like, without being prevented or pursued by any others except Us. . . . " (Article 19). Moreover, the last article granted the horologists permission to keep in their shops, in a public and visible place, a forge and a furnace to melt and forge all that pertained to their craft.

A legal decision, given on May 8, 1643, had already fixed the boundaries between the privileges of the horologists and those of the goldsmiths. The first could, in the words of this regulation, "make, sell & retail all kinds of cases, gold or silver, enameled & engraved," under the condition that "they will be able to buy gold & silver from only the said goldsmiths & no one else & that they shall work under the same standard that Master Goldsmiths have to respect."

Let us also note that watchmakers were obliged to write their names on their work and to stamp their mark, which had to be engraved on a brass table in the Registry of the Mint.

An article also protected the watchmakers from the fraudulent trade of the jeweler-haberdashers, about whom the eighteenth century watchmakers would so bitterly complain: "[It is ordered] that the merchant jeweler-haberdashers qualified to traffic in all kind of merchandise shall not be able to buy or sell any merchandise of horology in our city and suburbs of Paris, nor in other towns of our Kingdom, unless the said merchandise has been visited and marked, and found good by the said wardens of the said Horologist's art in our city of Paris, who shall be allowed to make visitations at the said merchant haberdasher-jewelers . . . in order to obviate the abuses and malfeasances which could be committed, to the great detriment of the public" (Article 14).

Three visitor-wardens, elected by the whole guild, were entrusted for a two-year duration to see to the enforcement of the statutes. In their inspections of guild members, they were assisted by bailiffs.

Not all the watchmakers practicing their art in Paris were constrained to respect the statutes of the guild. In the "privileged places," the former ecclesiastical enclosures, which were outside the jurisdiction of the Châtelet, horologists, like other craftsmen of varied trades, could set up a shop without proof of indentures and without a masterpiece. They worked freely, unimpeded by the visitor-wardens' supervision. But they had no right to receive work from the masters of the guild. Among the principal privileged places, were the Notre-Dame cloister and parvis, the Saint Benoît courtyard, the Temple courtyard, the Faubourg Saint-Antoine, the Saint-Jean-de-Latran cloister and the Saint-Germain-des-Prés enclosure. Watchmakers evinced a strong preference for the last two.[79]

The "Galerie du Bord de l'Eau" in the Louvre was also, in a different way, a very privileged place. There, the court watchmakers had their lodgings and shops, reserved since Henry IV for the most skilled artists in Paris. They were exempted from most of the obligations imposed upon the members of the guild. Several Parisian masters managed to obtain in the seventeenth century the enviable appointment of "horologist and manservant to the king."

The power of the Martinots, the longest dynasty of French watchmakers, asserted itself over the course of the century. Denis Martinot was from 1601 to 1611 the horologist to Henry IV then to Louis XIII. His son, also named Denis, succeeded in the position and in that of governor of the palace clock. His brother Zacharie succeeded him in 1637 as watchmaker to Louis XIII, and later to Louis XIV. Like the Martinots, the Bidaults enjoyed the favors of several kings. Claude I, who died in 1650, was watchmaker and manservant to kings Louis XIII and Louis XIV. Claude II was named king's watchmaker in 1628. Henri-Auguste (1629–1693) succeeded his father, Claude I, in 1652 as horologist to Louis XIV. The three masters had their lodgings in the Louvre. Antoine Ferrier passed on to his son Guillaume his appointment of watchmaker to Louis XIII in 1622, the year of his death; father and son also had lodgings in the Louvre. So did Antoine Femeritté, married to Marie Martinot, and watchmaker to the king between 1626 and 1629. As with the mastership, the position of horologist to the king was becoming a hereditary family possession.[80]

Though apprenticeship and a successful masterpiece were conditions essential to acquiring mastership, some watchmakers managed to escape those regulations by getting "*lettres de maîtrise*" (letters of mastership). They were granted by the kings on the occasion of happy events and permitted the beneficiary to enter the guild, as master, without undergoing any examination proving his aptitude to practice the trade. Needless to say, the masters in the community were not happy with such intrusion in their af-

11 Round watch. Case decorated with paintings on enamel. Movement signed "G Gamot A Paris." About 1650. D. 63; Th. 30.
The paintings of the case are inspired by the story of Theagenes and Chariclea; two are copied from paintings by Charles Poerson. The movement, equipped with a verge escapement, has a hairspring that was added later. The pawl is engraved and pierced with foliage and strawberry flowers. The openwork cock, with acanthus leaves, was probably changed at the time of this addition.
In 1645 Gamot was one of the masters who demanded new statutes for the Watchmakers' Guild of Paris.
Paris, Louvre, Olivier Coll., inv. OA 8318

12 Oval watch. silver and brass engraved case. Movement signed "Salomon Chesnon A Blois." First third of the seventeenth century. L. 43; W. 31.
The pawl and the one-footed cock are engraved and pierced with foliage and strawberry flowers. Salomon Chesnon (1572–c. 1634), married in 1604 and had a son named Salomon (1607–1683) who was also a master watchmaker in Blois.
Paris, Louvre, Garnier Coll., inv. OA 7025

fairs. They obtained in November 1652 the suppression of this privilege by letters patent from the king: "We say and order that in the future our edicts and *lettres de maîtrise* granted by grace of marriage, the birth of Children of France, coronations, royal entries in our cities, etc. shall not take place or effect for the said art of horology."[81]

This decree gives an interesting insight on the progress of Parisian watchmaking in the middle of the seventeenth century, and on the consideration the art enjoyed during that period.

First of all, the knowledge the art brings is praised: "Through the application of an unknown motion it makes one discover the degrees of the sun, the course of the moon, the effects of the stars, the disposition of the moments, the seconds, the minutes, the hours, the days, the weeks, the months and the years; the production of metals, the qualities of minerals . . . all the sciences contribute to the favorable success of these objects."

The text then celebrates the practical advantages of the horologic science: "The striking of a clock, carefully attended to, preserves the person of a patient from the distressing attacks of his pains, when the remedy is, in right proportions, given at the hour prescribed by the physician. . . . "; "a battle usually finds itself right at the point of its glory with the help of a precise alarm watch. . . ." Consequently, "the invention of the watch must indeed be considered the principal agent of rest, of men's sweetness and tranquillity."

Finally, the quality and the renown of Parisian watchmaking are underlined: "The masters up to now received in our said city have made themselves so skillful that their industry greatly surpasses that of the foreigners, as much in the beauty of their work as in the goodness that they particularly strove to keep in it . . . of which we derive an advantage of so great consequence that the most considerable persons in our court, the merchants and all our populations have lost the desire to look elsewhere for such work, and by means of this, the transport of our coin is now no longer done to faraway countries as formerly was the wont. . . . "

Blois

Under the reigns of Henry IV and Louis XIII, Blois became a true horological city. The workshops founded during the sixteenth century prospered, while several dozen new ones opened.

Abraham (1589–1671), son of Simon Gribelin, established himself in 1614. Salomon Chesnon *fils* (1607–1683) took over the management of his father's shop. Charles Peiras' son Salomon (1605–1684) was established by 1630. Michel Cuper (1590–1632), son of the founder of the dynasty, opened a shop in 1613. His nephew Barthélemy also opened his own the same year.[82]

The Blesian watchmakers continued to enjoy royal or princely favors during the first part of the seventeenth century. Several among them were named horologists either to the king or the queen. After being horologist to Queen Catherine de Médicis, Abraham de La Garde became horologist to King Henry IV. Nicolas Lemaindre (1598–c. 1652), obtained the same position from the queen, Marie de Médicis. Abraham Gribelin received the post under Louis XIII. Duke Gaston d'Orléans, the king's brother, showed favor to several watchmakers: Michel Cuper, Jean Locquet, Paul Rigault, Pierre Scander.[83]

In the inventory taken after the death of Mazarin (1602–1661) we found evidence of the success of Blesian watches. Out of the five watches belonging to the cardinal, four were the work of

13

14

13 Oval watch. Engraved silver and brass case. Movement signed "N Hubert A Rouen." First half of the seventeenth century. L. 65; W. 35; Th. 25.
Noël Hubert was the manager of *Le Gros Horloge* [*sic*] in Rouen between 1612 and 1650. In 1617 he, with other watchmakers, asked for the creation of a guild in Rouen. Several watches signed by him are still extant.
Paris, Musée National des Techniques (CNAM), inv. 19363

14 Round watch. Gadrooned silver case. Movement signed "G Seney Juré Pour le Roy A Rouen." Second quarter of the seventeenth century. D. 38; Th. 24.
Guillaume Seney participated, like Noël Hubert, with the watchmaker group that asked for a guild in Rouen.
Paris, Louvre, Olivier Coll., inv. OA 8286

15

15 Cruciform watch. Rock crystal case. Movement signed "Fonneveau a la Rochelle," first half of the seventeenth century. L. 57; W. 35.
Zacharie Fonneveau, a watchmaker active in La Rochelle, signed two cruciform watches now in the Petit-Palais.
Paris, Musée du Petit-Palais, Dutuit Coll., inv. 1413

Blesian craftsmen, Macé, Grégoire, Pierre le Roux and Jean Rou; the fifth was signed by a watchmaker from Saumur.[84] The inventory made in 1553 of the cardinal's furniture also mentioned a watch from Blois: "the movement of the diamond watch of the late Queen Mother made by Le Mindre from Blois, with its green and black enameled dial, in its gilt Turkish leather case," was valued at 600 livres.[85]

The technique of painting on enamel, perfected around 1630, was brilliantly used by Blesian goldsmiths in watchcase decoration. This success contributed to the rapid expansion of Blois watchmaking in the second third of the century and greatly influenced the specialization of Blesian shops in small-work timepieces. The enameled watches they produced went through an exceptional vogue, assuring them of an important clientele and preeminence in the French court.

In 1631 Nicolas Lemaindre received 2,135 pounds as "payment for having provided and delivered to His Majesty seven watches in gold cases, enameled with figures in all its parts."[86]

Contemporary references, collected by Abbé Develle, show to what extent the enameled watches had come to symbolize Blesian activity. The daughter of Gaston d'Orléans, Mademoiselle de Montpensier, notes in her *Mémoires* that as souvenirs of her visit to Blois in 1637, she brought back to Paris two watches as gifts to Louis XIII and Anne of Austria: "The one for the King was very small and enameled in blue; the one for the Queen was also enameled and had figures as was the fashion in these days. . . ."[87] When the aldermen of the city decided in 1643 to offer a gift to the duchess d'Orléans, Marguerite de Lorraine, they settled on "a watch in a gold enameled case with figures and personages."[88]

By 1660 the irreversible decline of watchmaking and other art industries in Blois was imminent. Several historical factors explain its causes: the death that year of Gaston Duke d'Orléans, which ended a long patronage; the concentration of arts and crafts activities in Paris under the personal reign of Louis XIV; and finally, more than anything, the religious persecutions against the Huguenots, which forced numerous art craftsmen from Blois into exile.

Lyons

The industry in Lyons knew its greatest prosperity during the first two thirds of the seventeenth century. At the end of Henry IV's reign, the city numbered sixteen watchmakers. This number rose to nineteen under Louis XIII and to twenty-six by midcentury.[89]

The craft remained unregulated in Lyons until 1660, when a guild was created. However, despite the absence of guild regulations, a number of principles were observed to ensure the serious professional instruction of the masters. Legal acts, rediscovered by Vial and Cote, such as indentures or journeymen contracts, outline the requirements. A future master had to have finished a five- or six-year apprenticeship, then to have served as a journeyman for about two years in a master's shop.

In 1658 the master watchmakers in Lyons found it useful to form a community. Their numbers, which had been steadily increasing, made them fear an evergrowing competition.[90] An assembly of twenty-eight watchmakers prepared its statutes, which were then approved by the city council. But the following July, twenty members retracted their support and wanted to rescind the projected statutes. They thought that the regulations were impractical and would bring more problems than advantages. They also feared that these regulations "would plunge them into constant lawsuits"; they were very good prophets on this point. The king had approved the proposed statutes, however, and the patent letters being published, they were registered by the Parlement on September 4, 1659. The new guild's regulations came into effect the following year.

The statutes of the Lyonese guild differed little from those applied in Paris. The most important difference concerned the mastership requirements. An apprentice wanting to establish himself as master watchmaker had to be a journeyman for three years, then work in a master's shop for three months. At the end of this short period, during which all his work had been checked, the candidate was judged capable or incapable of establishing himself as a master. The masterpiece, required in Paris, was thus replaced by a time of journeyman service, reduced to a year for the son of a master, and by a sort of continuous supervision.

Among the most renowned of the Lyonese watchmakers, active in the first two thirds of the seventeenth century, we may count Jean Blancheton, native of Saint-Etienne, who set up a shop in 1607; Sébastien Chappuis, who came from Geneva; Claude Champagnieu, who died in 1658; Louis Arthaud (1612–1662), famous for his astronomical watches; Jean Vallier, who was established as a

16 *Left.* Cruciform watch. Rock crystal case. Movement signed by Jean Rousseau. Geneva, second quarter of the seventeenth century. L. 43.5; W. 31.8.
Jean Rousseau (1606–1684) was apprenticed to his brother-in-law, Jacques Sermand. He had twelve children, two of whom were watchmakers.
Right. Star of David watch. Rock crystal case. Movement signed by Jacques Sermand. Geneva, first half of the seventeenth century. L. 31.5.
Jacques Sermand (1595–1651) is the author of several watches of various forms, now in the British Museum.
London, British Museum, inv. 74, 7–18, 28 and 88, 12–1, 193.

Jean
Rousseau
74
7-18
28.

master in 1602 and died in 1649; Hugues Combret (1596–1669); Guillaume Nourrisson (fl. 1645–c. 1700) and his brother Antoine (1628–1715), both appointed clockmakers to the city.[91]

Other watchmaking cities

Watchmaking prospered in many other cities in France. Rouen had enough watchmakers to start a guild. Its statutes, similar to those in Blois and Paris, were approved by patent letters dated July 1617. Important watchmaking families, such as the Huberts, worked in this city. The founder of that renowned dynasty was Noël Hubert, manager of *Le Gros Horloge* [*sic*] from 1612 to 1650. Another family, the Grébauvals, produced several watchmakers whose activity extended throughout the first half of the seventeenth century. A number of watches that have been preserved keep alive the memory of a few other watchmakers of Rouen: Guillaume Seney, who signed a watch in the Olivier Collection, also specifying his office: "King's Magistrate in Rouen" (inv. OA 8286), and Jean Thorelet, who made an oval watch now kept in the Louvre (inv. OA 683).

Several watchmakers practiced their trade in La Rochelle, among them Jean Flant, who, after a stay in Geneva, returned to establish himself in the city, and Zacharie Fonnereau (or Fonneveau), a master around 1640, the maker of a cruciform watch now in the Dutuit Collection (Petit-Palais, inv. 1413).

Isolated watches show proof of the development of the art in France in the first half of the seventeenth century. C. Phélizot was working in Dijon, as his signature indicates, on a watch now in the Musée International d'Horlogerie (inv. 485),[92] on one in the Mallett Collection and on still another in the Louvre (inv. MR.R. 241). Louis Guion also practiced in the same city.[93] Ballard was operating in Bourges, as is indicated by his signature on the movement of an oval watch.[94] Pierre Paris, who died in 1640, was established in Nantes; a watch kept in the Victoria and Albert Museum carries his signature (inv. 556–1901). These are but a few of the many examples of the vitality of French watchmaking during the first two thirds of the seventeenth century.

The development of two great watchmaking centers

During the first half of the seventeenth century, two other watchmaking centers destined for a brilliant future were developing: Geneva and London.

Geneva

On January 13, 1601, the Genevan watchmakers requested from their City Council the right to form their own community. Like their counterparts in Paris and Blois, they themselves composed the statutes they wanted to see applied. These were approved on January 20 by the Council.[95]

"Wishing their condition to be reformed through mastership, as it is in the good cities and towns in France,"[96] they had taken as a model the regulations of the French guilds. Apprenticeship was to last at least five years; masters could train only one apprentice at a time. Accession to the mastership was based on the success of two masterpieces, "a small alarm clock to wear on a chain around the neck and a square clock to keep on a table, with two heights." In other words, they meant a portable alarm clock and a two-level turret-shaped table clock. As in France, the statutes favored the sons of masters: they did not have to be journeymen for one year, as was required of other apprentices, before becoming candidates for mastership. The only masterpiece they had to make was a simple watch.

Geneva's vocation for watchmaking solidified through the first half of the seventeenth century. The watchmakers were now recruited from inside the city. A few written documents and the small number of watches we have from the period give us the names of a dozen watchmakers active at that time.

Martin Duboule (1583–1639) made a skull-shaped watch, now in the Paul Garnier Collection in the Louvre.[97] His son Jean-Baptiste (1615–1694) was a very talented master; the British Museum has one of his pieces, an astronomical watch of very fine quality. He also left his signature on a watch shaped like a lion, which belongs to the same museum.

Jean Rousseau (1606–1684), the famous philosopher's great-grandfather, is represented in several museums; he marked his signature on a skull-shaped watch in the Olivier Collection (inv. OA 8314), a rock crystal cruciform watch in the British Museum, a chased and engraved silver round watch belonging to the Victoria and Albert Museum, and a rock crystal round watch in the Garnier Collection. He had two sons, David and Jacques, who also became watchmakers.

Jacques Sermand (1595–1651) made most notably a rock crystal watch in the shape of a Star of David, now kept in the British Museum. His nephew Jacques Sermand (1636–1667) signed three astronomical watches, one of them in the Garnier Collection.

Few French watchmakers settled in Geneva during this period. Among the few who did was Antoine Arlaud, a native of Auvergne who arrived around 1617 and was to found an important watchmaking dynasty.

On the other hand, young Genevans went to learn the trade in French watchmaking cities and sometimes settled there. Martin

Duboule was a journeyman in Lyons around 1603. Fonnereau went to Lyons in 1618, then to La Rochelle, where he became a master in 1641. Pierre Bertrand set up shop in Lyons in the first quarter of the seventeenth century. Etienne Bouvier, born around 1642, also established in that city, where he renounced Protestantism in 1666.[98]

London

Over the course of the sixteenth century, the settlement of foreign craftsmen in the English capital was encouraged by Henry VIII, Edward VI, and then by Elizabeth I. Huguenot horologists asked for the benefit of the Crown's hospitality. The English court constituted indeed an important clientele, which was a major incitement to any watchmaker to establish himself in London. Henry owned in particular several clocks and sundials made by two immigrant masters, Vincent Quenay and Nicolas Urseau.[99]

As in Geneva, the rapid expansion of the horological art was promoted by the arrivals of French watchmakers. In the list of the Huguenots who obtained their citizenship on July 1, 1544, one finds eleven clockmakers, most of them of French origin, such as Robert Lambert, Charles Durant, Pierre Mare, Jean Demolyn, Jean Riche, Jean Votier, and Nicolas Urseau. The last was to obtain the appointment of horologist to King Edward. His son, also named Nicolas, had the same position with Queen Elizabeth.

Until the end of the sixteenth century, the industry in London was almost entirely in the hands of clockmakers and watchmakers who came from the continent. The scarcity both of written documents and of preserved watches does not permit us to measure the importance of watchmaking in the economic life of sixteenth-century London.

The oldest English watch kept in the British Museum, signed "Randolph Bull," is dated from 1590 (inv. 74, 7-18, 11).[100] From the beginning of the seventeenth century on, the number of English watchmakers kept growing. Several of the watches have been preserved. Their decorative style shows a strong French influence, due to the presence in the capital of numerous French craftsmen and the popularity of the watchcase designs being created by French engravers such as Etienne Delaulne and Antoine Jacquard. Many English watchmakers also ordered their watchcases directly from France, particularly from Blois. Thus, several watches made by David Ramsay (c. 1590–c. 1654) have Blesian cases.[101]

The first evidence of a joint action by the English horologists dates back to 1622. It was a petition signed by sixty watchmakers requesting measures against other watchmakers practicing the trade without permission. The petition was accompanied by a list of twenty-four names of watchmakers guilty of unfair competition. Most of them were Huguenots, natives of France. The request remained unanswered; a new petition was presented in 1627. The London masters opposed the freedom of trade of the French watchmakers.[102]

17 Oval watch. Movement signed by Randolph Bull. London, 1590. L. 57.6; W. 47.2.
Randolph Bull (fl. 1582–1617), governor of the Westminster Palace clock and clockmaker to the king, was one of the first native English watchmakers to succeed in the English Court.
London, British Museum, inv. 74. 7-18, 11.

Among these French watchmakers were many from Blois. Louis Cuper, the son of Paul, founder of the longest dynasty of watchmakers known, left the famous Blois family shop to establish himself in London in 1629. His nephew, Josias Cuper, also settled in London during the years 1620–1630. Pasquier Peyras, born in Blois in 1597 and a member of an important Blesian watchmaking family, was also working in England around 1632–1639.[103]

The London master watchmakers were at that time, in general, part of the Ironmongers' Guild. Their growth in number in the first third of the seventeenth century and the increasingly obvious difference between the refinement of watchmaker's technique and the informal character of the blacksmith's inspired them to de-

David Bußhman
Augustan

18 Round watch. Engraved gilt silver case. Movement signed David Bushman, Augsburg, second half of the seventeenth century. D. 50.
David Bushman (fl. 1640–1712) belongs to a family that brought Augsburg watchmaking to a high degree of perfection.
Vienna, Kunsthistorisches Museum, inv. 1564

mand their independence. On August 22, 1631, they received their charter, despite the opposition of the Ironmongers' Guild.[104]

The new community was directed by one master, three wardens, and "assistants," of whom there had to be at least ten. This managing committee formed the Court of Assistants. Admission to the Company was made by invitation and required the sponsorship of one of the members of the managing committee. The candidate was then admitted as a "freeman." Later, again upon invitation, the master could become a "liveryman," that is to say a full member of the guild; he was then entitled to be elected an assistant. The regulations did not force all watchmakers to be part of the guild. They were free to choose another guild if they wanted.

The statutes of the new guild, sixty-three in number, were approved on August 11, 1632.[105] Apprenticeship was to last at least seven years. Apprentices, after duly serving their masters and after their admission to the rank of freemen, had to work as journeymen for two years, and then to successfully produce a masterpiece before being admitted as "workmasters." The guild members could not employ any outsiders to the guild, whether Englishmen or foreigners. All imported clocks and watches had to pass an obligatory inspection before sale; in order to make sure the regulations were respected, the guildmaster and his wardens had the right to visit workshops and stores and to confiscate defective pieces and destroy them.

Other watchmaking centers

In Germany, the number of watches produced in Augsburg and Nuremberg was still substantial at the start of the seventeenth century. A census of the Augsburg population taken in 1615 gives the names of forty-three master watchmakers.[106] Another one, dated 1619, still gives the same numbers.[107] The Thirty Years' War was to cause the decline of the art industries in both cities. The extreme paucity of watches left to us that bear the signature of an Augsburg or Nuremberg horologist after the beginning of the seventeenth century confirms the almost total disappearance of watchmaking in both these cities. Germany will no longer ever figure in the first rank of watchmaking countries.

III From 1660 to 1730: The predominance of English watchmaking over the French and Swiss industries

Through the high quality of its technique, perfected by greatly talented masters such as Thomas Tompion, English horology asserted its primacy in the last quarter of the seventeenth century. On the other hand, French watchmaking underwent a decline in that period, provoked by various factors, such as the revocation of the Edict of Nantes, the general economic crisis at the end of Louis XIV's reign, the increasing rigidity of its guild regulations and the absence of master watchmakers comparable in skill to their English counterparts.

The decline of the French industry

Until around 1662, French horology underwent a brilliant expansion that left it peerless. This expansion was brutally interrupted by the emigration of a good number of Huguenot watchmakers and goldsmiths, who in turn gave a considerable boost to the industries of the countries in which they took refuge.

A political cause: the emigration of Huguenot watchmakers

The Huguenots were particularly active in business. The provinces and towns where trade and industry flourished contained a significant Protestant population. The art industries, particularly silk weaving, jewelry and watchmaking, counted a considerable number of craftsmen belonging to the Reformed Church. Protected in the first half of the century by the Edict of Nantes, which had been promulgated in 1598, they had been able to practice their trades under fairly good conditions and had contributed a great deal to the commercial and industrial development of towns like Blois, renowned for its jewelry and its horology; Lyons, famous for its fairs, its horology and its weaving; and Rouen and La Rochelle.

The repressive measures taken against the Protestants started with the personal reign of Louis XIV. In 1661 the king formed a commission to scrutinize the affairs of the Reformed Church. Persecutive measures followed: in 1667, the prohibition to perform burials in daytime; in 1669 the suppression of bipartite chambers of justice and a decree that at least half the members of a guild be Catholics; then came the interdiction to practice a number of occupations; after 1676, the creation of a convert's fund called the Pelisson Fund, followed the Dragonnades, furious persecutions by the military. These measures provoked a first wave of emigration in 1680 to 1683. The revocation of the Edict of Nantes on October 18, 1685, led to new departures abroad. Emigration picked up again after 1698, when Louis XIV reasserted the strict enforcement of his decrees.

The persecution before and after the revocation of the Edict forced the departure of at least two hundred thousand men, women and children. Such an exile would have had slight consequences for the economic life of France if these men had had no qualifications. But on the contrary, the Huguenots who chose exile belonged to the most industrious and most economically active class of Frenchmen.[108]

French horology numbered among its masters many followers of the Reformation. They had been playing an important role in the development of the industry from the mid-sixteenth century to the mid-seventeenth by taking part in the creation of great horological centers such as Blois and Lyons. Out of ninety Lyonese watchmakers, fifty were Catholics, while forty were Protestants.[109] The Huguenot masters established horology in cities favorable to the Reformation, such as Rouen, Sedan and La Rochelle. In these places, a genuine correlation could be made between the degree of development of the horological art and the proportion of Huguenots working there.

The Huguenot watchmakers migrated to the Netherlands, to England, to Geneva and to Germany. The departures for Holland were the most numerous, particularly for Amsterdam.

The large number of French Huguenots emigrating to Holland was a general phenomenon that did not concern watchmaking alone. Entire sections of cities such as The Hague, Rotterdam and Amsterdam were populated with Frenchmen.[110] Around 1715 the total number of Huguenots in Holland, was probably higher than that in any other country.

19 *Left. Oignon* watch. Smooth silver case. Movement signed "Marcou Amsterdam." Last quarter of the seventeenth century. D. 51; Th. 28.
The circular cock is engraved and pierced with a monogram under a ducal crown.
Michel Marcou, a Châtellerault native Huguenot, took refuge in Amsterdam in the last quarter of the seventeenth century.
Right. Round watch. Gold studded case. Movement signed by Thuret. Paris, end of the seventeenth century. D. 43; Th. 26.
A monogram decorates the case.
Isaac Thuret, a watchmaker to the king, died in 1706; he is the inventor of the movement characterized by not having a fusee.
Paris, Louvre, inv. OA 8415 and 7038

Several watchmakers from Sedan moved to Amsterdam: Jérémie Grandidier left for that city in 1681; Daniel de La Feuille, a master since 1669, migrated in November 1682. In Amsterdam, leaving his watchmaking shop to his son Jacques, he became an engraver and published in 1698 a *Livre nouveau et utile pour les horlogers, peintres et graveurs* . . . (A new and useful book for horologists, painters and engravers).[111]

After the revocation of the Edict of Nantes, watchmakers of various cities and towns of the kingdom settled in Amsterdam: Michel Marcou; Jacques and Guillaume Lucas, who had practiced in La Rochelle; Pierre Chesnon, master in 1672 and grandson of the great Blesian watchmaker Salomon Chesnon, who took refuge in 1686; Pierre Lemaire, master in Paris since 1675, who arrived in 1686; Jacques Mercier, who migrated in 1687; Charles Payras, a member of a Blesian watchmaking family, who settled in 1688; Nicolas Quesnay, who arrived from Dieppe in 1688; Noël Hubert's son Robert, who came from Rouen in 1689.[112]

In the archives of the city of Amsterdam, we found legal papers concerning the master horologists Jean de la Combe (an act dated 1701), Etienne Hubert the Younger (1703, 1704, 1710), Salomon Hubert (1709), Vincent Menil (1708), Isaac Ourry (1702, 1703).

Many Huguenots took refuge in England. The archbishop of Salisbury, Gilbert Burnet, wrote in 1725 that from forty to fifty thousand refugees had settled in his country.[113] The English government was far from hostile to the immigrants. Charles II made a declaration, published in the *London Gazette* of September 12, 1681, in which he promised that all the Protestant refugees would be most welcome and permitted to practice their arts and trades according to the laws of the kingdom.[114] These official guarantees were very useful to the immigrant craftsmen, who often met with the guilds' hostility. Among them were weavers, many goldsmiths,[115] jewelers and watchmakers.[116]

The last, for the most part, joined the Clockmakers' Company, whose numbers they reinforced. The exodus of the watchmakers had begun in the years prior to the Edict's revocation. They were not coming to a totally strange land, since many French watchmakers had already settled in the English capital during the first half of the seventeenth century. David Lestourgeon came to London from Rouen in 1681; his son, also named David, became a renowned watchmaker and was enrolled in the Clockmakers' Company from 1698 to 1731. Pierre Richard already belonged in 1679. James Gavelle joined from 1682 to 1700. Abraham Desessart was installed as a member in 1682. Thomas and James Terrier, established before 1685, were members of the Company as of 1694.

After 1685, the number of French watchmakers who settled in London multiplied, as evidenced by their increasing enrollment in the Clockmakers' Company. Pierre Debaufre, who became a master in Paris in 1675, is listed as a member of the Company from 1689 to 1722; this distinguished watchmaker, with the help of his brother Jacob and N. Fatio, was the first to make pierced rubies. Simon Beauvais, received as a master in Paris in 1670, a descendant of a Parisian watchmaking family, was a member of the Company from 1690 to 1730. Simon Decharmes, who surfaced in London in 1688, became a member from 1691 to 1730; his son, David, was much renowned for his watchmaking. Claude Duchesne, a Parisian, took refuge in London in 1689 and belonged to the Clockmakers' Company from 1693 to 1730. Other French watchmakers arrived in London in the beginning of the eighteenth century, among them Jacob Bouchet and Louis Gaudin.

The large number of horologists of French origin working in London during the eighteenth century is the result of several factors. First of all, the watchmakers coming from France often taught their craft to their sons. Furthermore, the descendants of the Huguenots numbered dozens of watchmakers; for instance, Gideon Perigal, who moved from Dieppe to London with his wife in 1688, had a son named Francis, who became a watchmaker and was elected master of the Company in 1756, and several grandsons who were also watchmakers.[117] And finally, the emigration of French watchmakers continued throughout the entire century.

Savary des Bruslons[118] was perhaps the first historian who underlined the importance of the French emigration in the development of English horology: "If the English make the case of disputing it [the perfecting of the art] with us, they owe it to the quantity of French watchmakers that the revocation of the Edict of Nantes obliged to take refuge in London; it would be easy to point out that three quarters of the watches that come from that country are made by Frenchmen. It would suffice to report their names, as well as those of the famous jewelers who, having left France for the same reasons, brought the perfection of the watch exterior to a point it would otherwise not have reached. . . . "

Between 1682 and 1700, about twenty-five thousand Huguenots[119] settled in Switzerland. Out of that number, Geneva welcomed three to four thousand. The city had received an earlier influx of French watchmakers before the Edict of Nantes (1598). We have seen that the coming of these immigrants had favored the establishment of watchmaking in Geneva. The revocation of the Edict brought the arrival of such watchmakers as Nicolas Bouquet, Paul Hubert, Jérémie Duchesne and Antoine Rey. But the second wave had much less impact on the horological activity of the city than the first. Genevan watchmaking was already self-sufficient. That is probably the reason that explains the exile of French watchmakers toward other parts of Europe where the industry had been more recently established.

The emigration of the Huguenot watchmakers is one of the fundamental causes of the ruin of the watchmaking industry in several provincial towns and cities, such as Blois, Lyons, Sedan and La Rochelle. But other disastrous factors played their part in the decline of the industry in France under the reign of Louis XIV.

Social and economic causes

The advent of the personal reign of Louis XIV signaled the start of the concentration in Paris of the art industries at the expense of the provincial cities. This concentration was promoted by the gathering in Versailles of the nobility of the kingdom, who formed the largest part of the mastercraftsmen's clientele.

The reign of Parisian horology began with that of the Sun King. But it took a span of two generations before it could assert its power and justify, through the quality of its masters, its preeminence.

The guild regulations tended to restrict the number of "new men" and to emphasize the privileges of the already established masters. Proof of the ossified condition of French watchmaking under Louis XIV may be found in the fact that the posts of horologist to the king had become the monopoly of two families.

The Martinots and the Bidaults handed down among themselves, like a hereditary possession, the privileges attached to the function of king's horologist: lodgings in the Louvre, relative independence from the guild, an annuity, and dinner at the table of the king's valets de chambre. The Martinots knew so well how to assert their claims that they held most of these positions in the last quarter of the seventeenth century and the first quarter of the eighteenth. Zacharie Martinot had the appointment from 1637 to 1684, his son Jean from 1664 to 1705 and his grandson Jérôme from 1689 to 1725. Gilles Martinot, Zacharie's brother, obtained one of the positions in 1662; Louis-Henri, his son, got one two years later; Balthazar Martinot (1636–1712) was horologist to Anne of Austria and then to Louis XIV; his brother, Claude, received the same charge. The celebrity of the name Bidault rests on the number of appointments as king's horologist that the family succeeded in accumulating. Claude Bidault I was horologist to Louis XIII, then to Louis XIV. Henri-Auguste succeeded his father in 1652; the son of the latter, Auguste-François, was in turn named horologist to the king in 1693. Augustin-Henry Bidault received the charge in 1705.[120]

While in England the watchmakers multiplied their efforts and their inventions to improve the running of watches and clocks, in France very few watchmakers, if any, elevated to the level of their researches.

Until the time of the Regency, horological research in France was conducted by scientists. Studies pertaining to horology, presented to the Académie Royale des Sciences, were undertaken by scientists such as Philippe de La Hire, Abbé Varignon, Christiaan Huygens and Leibniz. The French watchmakers were only craftsmen. Isaac Thuret, the king's horologist, was thus hired by C. Huygens in January 1675 to build the first watch with a balance spring. In November 1692 the watchmakers Balthazar and Gilles Martinot, Nicolas Gribelin and Jacques Langlois contracted with Abbé de Hautefeuille to use an invention of his, aimed at improving the precision in watches.[121]

This deficiency of French horology had the clear result of giving the advantage to the English competition. A January 1719 article in the *Mercure françois* underlined the wrongs caused by this competition: "It is a fact known by everyone that every year England provides France with a large quantity of high-priced Watches, & that this trade only means a pure loss for France; these kinds of Pieces being ordinarily bought on commission from the most famous Master Watchmakers of London & paid for with ready money. Sheltered by the reputation the English have so rightly acquired, through the genius & industry they have made apparent by perfecting Horology & through the goodness & the neatness of the products that leave the hands of their best Masters, it has come to pass that a good many people of bad faith abuse the Public. Several English Watchmakers are only sending to Foreign Countries their Waste Goods. These in turn have been copied in Geneva, in Germany & in Holland. All these countries in Europe are being flooded with these bad watches; they can be compared to false Coinage spread among the public, to its great prejudice. The French Watchmakers have so much suffered of the abuse that their trade is almost ruined." The prestige of English watchmaking had thus provoked the appearance of a double competition: one fair, due to the quality of watches made by the English masters, and the other fraudulent, the result of poor-quality watches, claimed to be English-made, swamping the market.

The situation of French watchmaking was made more difficult by this trafficking in foreign watches: "It is the necessary consequence of what we have just said that the French Watchmakers presently have no outlet in Foreign Countries for their Watches; & what is worse, they cannot even provide inside the Kingdom without having to draw from London or Geneva some of the things they need to assemble a Watch."[122] The author of the article was hinting that French watchmakers imported unfinished movements and watchcases from London or Geneva. Indeed, under the reign of Louis XIV, Genevan competition had started to assert itself. Throughout the eighteenth century, it will be denounced as a real calamity by the French watchmakers.

Other, more general causes harmed the vigor of the industry in France. Thus, the grave economic crisis that marked the end of Louis XIV's reign, caused by the disasters of the war, by the famine of 1709–1710, and by the ruin of the treasury, was highly prejudicial to the development of luxury industries such as watchmaking. Sumptuary laws promulgated at the time put a brake on the production of luxury objects, and consequently on that of watches.

Under the Regency: a new awareness of the weaknesses of French watchmaking

The first concrete evidence of an awareness of weakness was the establishment of a watch factory in Versailles. The project for that establishment was conceived in February 1718 by an English watchmaker, Henry Sully. He called on Law, the general comptroller of the Treasury, to submit his proposal to the regent. Philippe d'Orléans approved the project and charged Law with helping Sully in the founding of the factory. A decision was made to establish it in Versailles and to call on the services of English watchmakers.

According to the January 1719 article from the *Mercure de France*, entitled "Relation abrégée de l'établissement de la nouvelle fabrique d'horlogerie à Versailles" (A shortened account of the establishment of the new watch factory in Versailles): "An assortment of the most excellent English workers in each branch of Horology was brought, at great expense, to France. They were settled in *Versailles* in Royal Houses and given considerable advantages. . . . " So that the establishment might serve to perfect French horology, a school was created: "Young people are admitted to learn Horology in all its parts; & they are taught the Theory as

20

21

well as the practice of their Art. This establishment is thus at the same time both a factory able at present to produce works of the utmost beauty & perfection, & an academy able to turn out skillful workers & scientists for the future."

The aim of the factory was to return French horology to its former splendor. The same article stresses the idea in these words: "The design of this Establishment is to . . . raise again the honor of Horology in France, to enable it to attend to the needs of the Kingdom without any foreign help & be able to supply the outside trade."

When the first watch made in the factory was presented to the King on January 17, 1719, everything indicated that the factory, under Sully's management, would fulfill its mission. On the same day, the regent received a watch with repeater, also made by the watchmakers of the factory, and he was assured that the kingdom could anticipate considerable advantages from this enterprise.

20 *Left. Oignon* watch. Engraved silver case. Movement signed "Decharmes London." End of the seventeenth century. D. 49; Th. 32.
Simon Decharmes, a Huguenot refugee in London, belonged to the Clockmakers' Company from 1691 to 1730; he is the inventor of this alarm movement, which is decorated with engraved acanthus leaves.
Right. Oignon watch. Gold-studded leather case. Movement signed "Baltazar Martinot A Paris." End of the seventeenth century. D. 62; Th. 38.
The movement has a mock pendulum balance that can be seen under the circular cock, which is decorated with foliage. Watches equipped with this "fantasy" were very much in vogue toward the end of the century.
Balthazar Martinot (1636–1716) was a renowned watchmaker. After being one of the queen's watchmakers, he was appointed a watchmaker to the king's Council.
La Chaux-de-Fonds, Musée International d'Horlogerie, inv. 463 and 448

21 *Left.* Round watch. Engraved and pierced silver pair case. Movement signed "G Graham London" and numbered 741. Around 1735. D. 67; Th. 36.
The famous watchmaker George Graham (1673–1751) introduced in this movement the cylinder escapement he perfected in 1725. The balance with the regulating spiral spring is protected by a cock with a large triangular foot, which is engraved with foliage. The pivot bearing of the balance wheel tree is jeweled. A numbered rosette allows one to regulate the length of the hairspring. The movement has a repeater strike and an alarm.
Right. Round watch. Painted enamel case. Movement signed "Tompion London." End of the seventeenth century. D. 49; Th. 33.
The movement is characteristic of the production of Thomas Tompion (1638–1713), a pioneer in horological precision engineering. The winding square is sited on the plate. A numbered regulating rosette permits the modification of the balance spring. The balance is protected by a large cock, held by a foot engraved and pierced with foliage scrollwork.
La Chaux-de-Fonds, Musée International d'Horlogerie, inv. 398 and 397

Unfortunately, even before the end of 1719, the factory was in its death throes. In 1720, with Law in flight and an economic crisis beginning, it had to close its doors. The causes of the bankruptcy were variously analyzed. According to Savary des Bruslons, "the factory fell of its own weight before the year had passed, solely because of the French prejudice that these Watches were not made in England"![123]

Philippe d'Orléans' taste for French watches, but also for English, is apparent in the inventory of his estate done on April 28, 1724. It mentions eight watches, two with repeaters. Two had been made by French masters, Gaudron and Martinot; two others came from England. The repeater watch by Gaudron had an engraved gold case. Martinot's repeater was the most splendidly decorated; his gold case was "garnished with several small garnets and diamonds." One of the English watches was thus described: "with two gold cases with its golden chain and a steel hook." Four watches had shagreened cases.[124]

A little later, after abandoning the management of the Versailles watchworks, Sully was appointed by Duke de Noailles, the director of a factory established in Saint-Germain. This one, too, had to close in 1720. The discontent of the English watchmakers, who probably were disturbed to see their fellow countrymen working for the good of the French, hastened its closing. We read in the preface for the horological plates of the *Encyclopédie*: "That factory had been too well conceived for English jealousy to let it live very long. Soon England recalled her subjects. Most returned home & only left behind them the emulation set off by competition."

The vitality of English watchmaking

From 1660 to 1730, English horology asserted an incontestable dominance over that of the Continent. The revolutionary inventions by the scientist Christiaan Huygens that made possible the conquest of precision, the pendulum (1657) and the balance spring (1675), formed the point of departure for the brilliant development that horological technology was about to undergo in England.

The upswing of English watchmaking was determined by favorable economic conditions, but also, and above all, by a galaxy of remarkable watchmakers.

Favorable conditions

The Stuart Restoration in 1660 was a period of economic prosperity. London became at that time an international trading center. To the riches brought by trading in the East Indies, West Africa and North America were added the profits made in the rising industrial trades (shipbuilding and wool and cotton manufactures). The appearance of a rich merchant middle class in London society and the constant flow of tradesmen to the English capital increased the clientele for craftsmen, particularly watchmakers.

The influx of a large number of qualified French workers also favorably affected London's economic life. It can be stated, for instance, that the English silk industry developed with the arrival of the Huguenots. As we underlined before, the Clockmakers' Company of London was reinforced by the immigration of many French watchmakers at the end of the seventeenth century. The result was an increased production of horological pieces, notably watches; but this increase was also a response to a strong local and foreign demand.

A galaxy of great watchmakers

It was in England that the conquest of precision—henceforth possible thanks to Huygens' inventions—was undertaken. Around 1680 the supremacy of English horology had become incontestable. That preeminence was due to the remarkable talents of several masters.

Thomas Tompion (1638–1713), honored with the title "father of English watchmaking," may be considered the pioneer of the modern watch. In the domain of small-work horology, he was the first to foresee all the improvements the invention of the balance spring made possible. He conceived of a new model of watch, which became the prototype of the eighteenth-century English watch. The use of a pair case became an absolute rule, a large cock maintained by a foot protected the balance, the winding up of the movement was done on the side of the plate, and a regulating rosette allowed the lengthening or shortening of the regulating spiral spring. Responding to the demand of a vast clientele, Tompion set up a shop that produced a large quantity of watches built according to the model he had created. During the reigns of Charles II, James II and William III, Tompion was the premier court horologist. Enjoying great public respect, he was granted the privilege of burial in Westminster Abbey.

Accepted in 1671, like Thomas Tompion, in the Clockmakers' Company, Daniel Quare (1648?–1714) rose to the rank of the best English watchmakers. Because of his religious convictions, which did not permit him to take the oath of allegiance to the Anglican Church, he refused the appointment of horologist to the king when it was proposed by George I. Small-work horology owes to him the invention of the repeater mechanism, which had a great vogue in the eighteenth century. Quare was one of the first to use the "dust-cup," which protected the movement of the watch. He was also one of the first to introduce the minute hands in watches.

After 1700, English horology maintained the dominance it had acquired thanks to Thomas Tompion. New discoveries, such as pierced rubies, kept up the quality of English-made watches. English masters enjoyed an immense reputation on the Continent. The prestige of watches with the signature of Tompion, Quare or Graham was such that unscrupulous tradesmen could become rich by having watches manufactured in Holland or Switzerland falsely signed with one of their names.

George Graham (1675?–1751) contributed, like his master Thomas Tompion, to the hegemony of English horology. After his admission in the Clockmakers' Company, in 1695, Graham joined the establishment of Tompion, whose niece he married. This collaboration, which lasted until Tompion's death in 1713, obviously proved successful, since in 1711 Graham became the illustrious watchmaker's partner. Tompion bequeathed him his workshop under the sign of the Dial and the Three Crowns in Fleet Street. In 1722 Graham was named a member of the Royal Society Council and a master in the Clockmakers' Company. These honors were a reward for the quality of his work and his inventions. In 1725 he perfected the cylinder escapement, which he introduced, starting in 1726, in every one of his watches. In 1726 he invented the mercury compensated pendulum. His watch production was considerable; according to Britten, he made close to three thousand of them. As a last testimonial of its gratitude, England gave him the honor of being buried in Westminster Abbey, next to Thomas Tompion.[125]

The Genevan Factory

In the second half of the seventeenth century, horology asserted its dominance in Geneva's economic life. Because of the number of watchmaking shops and those devoted to complementary activities, the city seemed like a gigantic clock and watch factory. Its regulation, organization and markets became more precise over the course of the period.

Regulation and organization of production

Like their counterparts in Paris, the Genevan masters were anxious to increase their privileges and to protect themselves from competition. On November 28, 1673, their guild passed a new system of regulations that addressed these preoccupations.[126]

The conditions for admission to mastership became stricter and tended to limit even more the number of masters. The masterpiece, a portable alarm clock, had to be executed in the shop of one of the four jurors entrusted with supervising the regulations. Many residents were excluded from mastership, which was henceforth reserved to *citoyens* and *burghers*. On the other hand, the statutes favored the sons of masters; for them, the masterpiece was only a simple watch, with no complications.

On March 10, 1690, a new ordinance was approved by the Council of the Two Hundred. It accentuated still more the rigidity of the conditions required for mastership. Candidates were henceforth to be at least twenty-two years old. It was not only impossible for the *habitants* and the *natifs* to become masters, but they were also forbidden to start an apprenticeship in watchmaking.

The regulation was also aimed at preserving the Genevan masters from the competition that was starting to appear in the surrounding countries. In particular, it was forbidden to buy and sell watches made in the regions adjoining Geneva. The masters also mistrusted female competition. The statutes firmly excluded women from mastership. Female labor was permitted only in the fabrication of chains and the polishing, cutting and piercing of a few engraved parts of the movement, such as the cock. The women specializing in this kind of work were called the *vuideuses*.[127]

The numerical power of the guild of Genevan Horologers, was already considerable by the end of the seventeenth century. In his *Histoire de Genève*, published in Amsterdam in 1686, an Italian refugee named Gregorio Leti indicated that there were in the city more than a hundred masters and three hundred journeymen.[128] These horologists, according to the same writer, produced five thousand watches yearly.

The division of labor, organized in the second half of the seventeenth century, contributed greatly to the increase of production. Task distribution was divided in three parts: the preparation of the rough movements, called *ébauches* or *blancs roulants* (roughcasts or, literally, rolling blanks); the finishing of the movements; and the fabrication and decoration of the cases. The making of the roughs was given to the journeymen and, despite the regulations, to craftsmen in neighboring regions. The masters kept for themselves the finishing, which determined, more than any other work, the quality of the watch. The increased production provoked the appearance of specialized workers: springmakers, chainmakers, keymakers, etc.

The introduction of horology in Geneva had brought with it the development of the manufacturing of watchcases. In the city, which already counted numerous goldsmiths and jewelers, this new activity appeared a windfall. Several goldsmiths specialized in it, while new craftsmen, in larger numbers, devoted themselves freely to the new fabrication. The growing importance of that specialization ended up with the creation of the Case Assemblers' Guild. The "Regulations and Ordinances on the State of the Master Assemblers of Watch Boxes and Cases" were approved on August 16, 1698.[129] Apprenticeship was to last four years. Accession to a mastership was based on the success of a masterpiece, "a box carrying a case and striking box." The challenge was, in other words, a watchcase protected by a second case and a pierced case that was to contain the striking movement.

In 1716 the engravers specializing in watchcase decoration were numerous enough to obtain the right, despite the opposition of the goldsmiths, to form their own guild. The *habitants* were excluded from the new community. Apprenticeship was for five years, journeymanship for two. A candidate for mastership had to successfully make "a case or a box for a striking movement engraved everywhere, in *taille-douce* or embossed or in other work."[130]

The Factory's markets

From its beginnings, local demand being negligible, the Genevan industry was dedicated to exporting its products. The conquest of each new market, in Europe or the East, marked a new stage in its expansion.

The development of this outside trade caused the formation, in the second half of the seventeenth century, of a class of merchants specialized in the sale of horological goods, recruited from among the masters of the craft. Taking advantage both of their knowledge of the markets and of the absolute necessity for the clock- and watchmakers to dispose of their merchandise abroad, these merchants, called *établisseurs*, did not have the simple role of sellers. They also supervised the work of the horologists and artisans participating in the manufacture of watches.

A. Babel distinguished three main exportation techniques: the sale by *établisseurs* and their agents at fairs or during visits to the shopkeepers in large towns; the overseas selling by brokers; and the supplying of Genevan traders settled in faraway cities like Constantinople. These diverse selling procedures were well adapted to the diversity of the regions where, from the seventeenth century onward, Swiss watchmakers carried on their commercial activity.

To increase the number of their customers in Europe, the *établisseurs* assiduously frequented the large fairs, such as the ones in Lyons, Frankfurt, Basel and Leipzig, or they visited the tradesmen in Paris, Amsterdam and London. The exportation of Genevan watches to France was already quite important at the start of the eighteenth century.

The introduction of Swiss watches in the Turkish Empire was facilitated by the establishment in Constantinople, in the early seventeenth century, of a colony of Genevan watchmakers. The orders from the watchmakers of that city to the watchmakers of

Geneva were numerous. In April 1671 the watch merchant Sébastien Chappuis ordered fifty astronomical watches from Abraham Arlaud and forty others from Jean-Antoine de Choudens.[131] In the early eighteenth century, Isaac Rousseau, the father of Jean-Jacques Rousseau, embarked for Constantinople, where he was to repair Geneva-made watches; not having forgotten this adventure, his son attributed to him the glorious function of watchmaker of the Seraglio.[132]

IV From 1730 to 1820: Three great watchmaking countries—France, Switzerland and England

Over the course of the last two-thirds of the eighteenth century and the start of the nineteenth, a balance seems to have been established between the three watchmaking powers. The strength of Swiss horology at that time appears to have been founded on the prosperity of its trade with numerous markets throughout the world, while the prestige of both its English and French counterparts depended mostly on the production of watches of a high technical and artistic quality, reserved for an elite clientele.

The revival of French watchmaking

A rich clientele in love with refinement and science

The conditions that determined the extraordinary development of French horology from the 1720s onward deserve a few remarks. The author of the preface describing the horological plates in Diderot's *Encyclopédie* expresses this thought about the situation of French watchmaking around the mid-eighteenth century: "[It] has made itself so superior in the past forty years that it has acquired the highest reputation even abroad, where it is nowadays preferred to any other, because it dominates above all through its good quality & its taste."

The birth of so many horological masterpieces and the discovery of many technical improvements are tied to the social history of France in the eighteenth century and to the mentality of a clientele enamored of both refinement and science.

Luxury in the eighteenth century took on a new meaning, different from the ostentation of old. The latter sought to dazzle and to subjugate others; in the eighteenth century, luxury became a more intimate satisfaction, less showy, "a well-being and a comfort entirely material, where the object offers itself for enjoyment."[133] It was organized around the individual to give him more a personal enjoyment than a reflected glory. Objects used in daily life took on importance; clothing, jewelry, furniture and trinkets were indispensable to the life-style of the well-to-do of the period.

The nobility held a large part of the kingdom's wealth, and rare were those who invested it in land or industry. Most of them were content to spend it in acquiring luxury objects. This clientele favored the development of the bourgeois class containing the master craftsmen and the tradesmen, who provided them with the luxury items necessary for the ornamentation of their existence. In his historical study on the eighteenth century, R. Mandrou rightly notes that "crafts and commerce, bound to luxury consumption, have not ceased to benefit from the changing of fashions, needs and their exigencies for the wealthy population of the cities."[134]

In addition to this taste for luxury, the scientific curiosity of the century certainly contributed to the expansion of the horological trade and, more generally, to the development of the mechanical arts. Large passages in Diderot's *Encyclopédie* show evidence of the new interest brought to bear on the works of craftsmen. The deliberate intention to grant the mechanical arts an important place in a dictionary whose objective was to expose to the world all the knowledge of the time had been asserted by Diderot in the text of the *Prospectus* (1750), and stressed again by d'Alembert in the *Discours Préliminaire*: "Too much has been written on the sciences; not enough on most of the liberal arts; almost nothing on the mechanical arts." D'Alembert added apropos of horology: "And not to leave the domain of horology, why are not those to whom we owe the watch fusee, the escapement and the repeater, esteemed as much as those who worked in succession to perfect algebra?"

The watchmakers were in those days a privileged category of craftsmen, for their works, relating both to art and to science, corresponded perfectly to the tastes of a society enamored of luxury and technical novelties. They found a situation propitious to the expansion of their art, thanks to a clientele attracted by the complex mechanisms of the movement and by the precious decoration of its casing, which, itself an object of delight, often took as the subject of its decoration the pleasures of the period. This clientele, much wider than in past centuries, grouped nobles and

22 Parisian wearing a watch on a chain hanging down her shoulder. Plate 2214 in the 1824 issue of *Le Journal des dames et des modes.*
Paris, Musée de la Mode et du Costume

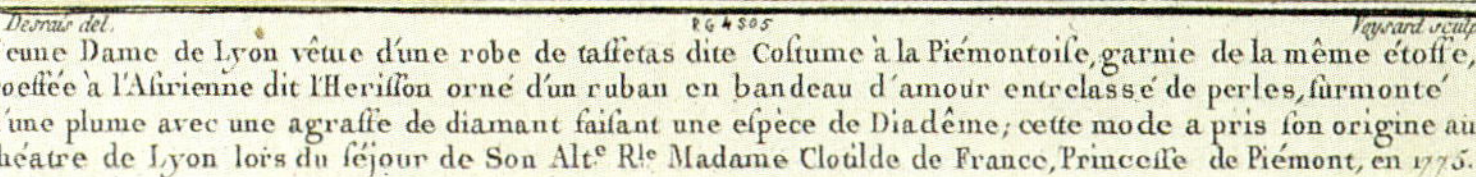

23 Fashionable types wearing watches. Paris, 1778 and 1779. Drawings by Desrais.
Left. Engraving by Voysard, plate N 78; 13th sketchbook of French costumes, 7th fashionable dress series of 1778.
Right. Engraving by Dupin, plate S 108; 18th sketchbook of French costumes, 12th fashionable dress series of 1779.
Paris, Musée de la Mode et du Costume

bourgeois—a certain type of bourgeois, those possessing a private income or an office.

In the eighteenth century, the watch became an instrument of practical utility that aroused scientific curiosity and at the same time a very elegant accessory to fashionable attire. It was good form for the men and women who cared about their elegance to wear two watches, perhaps to verify the exactness of one against the other. Prints showing the fashions of the time are evidence of this: they often show female or male models wearing two watches hanging at their waist by chatelaines. If his *Memoirs* are to be believed, Casanova was a follower of that fashion: "In Paris, to an invitation, I wore violet satin breeches and an ash-gray coat, the ruffles of which alone were worth a thousand livres. I displayed on my chest the cross of the order; last, I took two watches and two richly chased snuffboxes."[135] In the time of the Directory, an elegant woman wore her watch hanging from her neck by a chain or a very long string.

The French king was very interested in watchmaking; like the snuffboxes to which they were related by their decoration, watches were present in the list of "Royal Gifts." Royal or princely weddings and ambassadors' visits were occasions for him to present richly ornamented watches.

Louis XIV had instituted this custom, as shown in the document "Compilation of the Presents Offered by the King in Gems, Furniture, Silverware and Others, Starting in 1669, Until and Including the Year 1714."[136] On May 20, 1681, for instance, he gave sixteen watches to the ambassadors of Muscovy. An order of payment, found in the Archives Nationales, gives a few details about the presents and mentions the name of the watchmaker from whom they were ordered: Baronneau. The document reads: "To Baronneau, horologist, 2,489 livres for two clocks and sixteen watches which were given, to wit, to the first ambassador one clock and six watches, to his son, four watches and to the Chancellor, one clock and six watches."[137]

The tradition of offering horological pieces continued during the Regency. On August 1, 1721, the Turkish ambassador received from the king "a clock with its case in gilt bronze" (1,500 livres), his son "a gold watch with two diamonds" (1,600 livres), the master of ceremonies "a gold watch costing 800 livres," the physician "a repeater watch with its gold string" (910 livres).[138]

On May 16, 1770, at her wedding to the dauphin, Marie-Antoinette, archduchess of Austria, received a pannier holding fifty-two snuffboxes and fifty-one watches, which she distributed to the guests.[139]

Louis XV seems to have followed with great attention the progress of horology and the debates it kept raising. He was interested in particular in a quarrel between Caron *fils*, later known under the name of Beaumarchais, and Lepaute, his official horologist. The Académie des Sciences sided with Pierre Augustin Caron, recognizing him as the true inventor of a double-virgule escapement, which Lepaute had dishonestly attributed to himself.[140] The king wanted to meet the young victor; when Caron came to Versailles, the king ordered a watch for himself, another for Madame de Pompadour, and a clock for one of his daughters, Madame Victoire. Caron worked zealously and, without any delay, brought the royal watch to the king's grand levee. The king wanted him to explain the mechanism in front of all the lords present. A little later, Caron presented the watch for Madame de Pompadour. Its conception was entirely novel, as much for its escapement as for its smallness.

In a letter addressed to the *Mercure de France* on June 16, 1755, Beaumarchais announced his creation: "I had the honor of presenting to Madame de Pompadour a few days ago a watch mounted on a ring[;] . . . it is only four lignes in diameter and one ligne less a third between the plates. To make this ring-watch more convenient, I devised instead of a key a circle around the dial, with a small protruding hook. By pulling this hook about two-thirds around the dial with a fingernail, the ring is wound up and runs for thirty hours. Before bringing it to Madame de Pompadour, I saw this watch exactly follow my seconds-counting clock for five days; thus, by using my escapement and my construction, one can make excellent watches as flat and small as one wishes." Louis XV recognized the young watchmaker's ingenuity and made him his horologist.

In his book of accounts, Lazare Duvaux, a merchant jeweler, bears witness to the court's infatuation with watches. Watch prices were very high. In 1755 Marquise de Pompadour paid Lazare Duvaux 1,500 livres for an enameled gold watch with matching chain, containing a movement by the horologist Etienne Le Noir.[141] In the same year, she took delivery of a simple-movement watch in a gold-decorated porcelain case, with a steel chain, an agate seal in the shape of a dog *(cachet en chien d'agathe)* and a gold pair case from England[142] for 372 livres.

Kept from 1779 to 1793, the *Journal des ventes* of Ferdinand Berthoud's shop lists about fifteen hundred watches.[143] The document is of very great interest because it shows cost and sale prices. Thus, an equation watch with repeater, indicating the seconds (no. 1187), was delivered to Madame Adélaïde de France on September 21, 1783, for the sum of 2,400 livres. Its original cost was estimated at 1,228 livres.

The profit taken by that shop was usually less high and amounted to about 30 to 40 percent of the cost. A simple cylinder watch (no. 2083), engraved, was sold for 500 livres on September 17, 1784; its original cost put at 376 livres. A simple cylinder watch, with a blue enameled bottom, valued at 372 livres, was delivered to Viscountess de Satouche for 600 livres on May 25, 1784 (no. 2067). A simple verge escapement watch (no. 2031) was delivered for 348 livres; its cost had been 305 livres. A repeater watch indicating the seconds (no. 2135) was sold for 1,200 livres to M. Maslet on August 9, 1785; its cost had been 905 livres. A repeater and verge escapement watch was sold on June 4, 1784, for 720 livres to M. de Montalei; its cost had been 484 livres.

Louis XVI, whose taste for the mechanical arts is well known, owned many watches and clocks, in which he demanded unfailing precision. The horologist Caranda, valet of the king, in charge of the royal timepieces, demanded a raise, because he was "obliged to be with the King every day at the hour of his levee, to be on the Meridian and make the clocks agree again. . . . "[144]

In his private workroom in Versailles, Louis XVI used "a watch stand with five gilt brass hooks having in the middle a medallion with five fleurs-de-lys." For the Tuileries Palace he ordered from Robin an elegant watch stand, which was executed by the famous cabinetmaker Riesener. The bill of delivery written by Robin described it thus: "Tuileries Castle—delivered a watch stand in which to shut in the watches of the King. The said mahogany wood

Art. 3. Suite des Montres vendues depuis Juillet 1783

N°. 2044. Montre à secondes

Dépenses du mouvement estimé	96		
Finissage par Mr. Lallemand	132		
Boëte N°. 245 de Sanguinède	136	12	
Remontage doré & fourniture par Mr. Allier suivant le détail ci après	120	10	
Remontage . . . 60.			
Emboëtage . . . 15.			
Ressort de Cadran . . . 4.			
Dorure . . . 6.12.			
Gravure de bâte, rosette & nom . . . 6.10.			
Trou de Cadran . . . 1.4.			
Aiguilles . . . 24.			
Cristal . . . 1.			
Poli de Boëte . . . 1.4.			
Goutte de grande roue . . . 1.			
120.10.			
Gravure par Cassin	24		
Etui	16	10	
Clef d'or	10		
Spiral	6		
Cadran	9		
550.12.	550	12	

1783 xbre 16 Livré à Mr. Desbarolles pr. 730. a déd. 5 pr. de Remise accordée par le défunt reste 693.10.
Reçu le d. jour

24 Excerpt from the *Journal des ventes* of the shop of Ferdinand Berthoud (1727–1807), established in Paris, rue de Harlay.
The account gives details about the price and the execution of each part of a seconds watch, carrying the number 2044, delivered on December 16, 1783. Payment was 683 livres, discounting "the remittance granted by the deceased" (Henri Berthoud).
Paris, archives of the Musée National des Techniques (CNAM)

watch stand, garnished with moldings, and the frame as well, chased in matte gold; the door opening with a pushbutton and decorated with a mirror and, crowning the whole, the arms of the King; all the mahogany wood, polished and glossy like the mirror." The piece is now kept in the Château de Versailles Museum.[145]

The organization of production

The interest that eighteenth-century society took in watchmaking manifested itself in the granting of new statutes favoring the horologists; in a deep respect for the profession, whose greatest masters were protected by the court; and by the creation of premier horological workshops, meant to satisfy the demands of an ever larger clientele.

THE REGULATIONS OF THE PARISIAN GUILD

The Corporation of Horologers of Paris received new statutes in 1726, completing those approved in 1646. The new regulations were applied without any changes until the end of the century. The twenty articles that formed the new regulations outlined the conditions for admission to mastership and the duties of the jurors in charge.[146]

The apprentice who wished to become a master had to execute successfully an alarm or a repeater watch: "We permit Apprentices who shall have the required and necessary qualities to attain to the Mastership, to be received as Masters, by making a Masterpiece with an alarm or repeater clock, producing its effect in its case, at the choice of the Aspirant; which Masterpiece shall be handed back to him after it will have been judged perfect. . . . " (Article II). The candidate had to pay three livres for the royal fee, two hundred for the Community Chest and another three for the fee of the clerk. He had also to pay the cost of the letter of mastership and come forth with a "Purse of silver tokens weighing three marks," which were distributed to the masters forming the jury. Masters' sons paid only the royal fee, thirty sous to each juror in charge and the expenses of the letter of mastership.

The new regulations authoritatively reminded the jurors of their obligations regarding the enforcement of the guild's statutes. Article IX forbade them "to receive any Masters of the said Trade, either Sons of Masters, Apprentices, or others, under any pretext whatsoever, who have not already made a Masterpiece. . . . " According to Article X, they were obliged "to take an oath before our Royal Prosecutor in the Châtelet, as to whether the Aspirant has made and perfected the Masterpiece, and completed his Indentures, under pain for the Jurors of destitution and of not being able ever again to be counted among the Elders. . . . "

Despite its severity, the regulation did not prevent abuses. The fees that the masters had to pay on the occasion of their admission offered the jurors the temptation of easy money. A case memorandum, printed in 1738, underlined the existence of such malfeasance:[147] "According to the statutes of the month of March 1645, Article VII, the number [of horologists] is restricted and limited to seventy-two, and at this time there are close to three hundred. If the behavior of the Juror-Warden Horologists were examined in this respect, it would be found that they drew very considerable sums of money, which they applied to their profit, in order to receive Masters, not only unqualified Workers but even Genevan Foreigners. . . . " The passage gives us, moreover, interesting information on the number of Parisian masters in the first half of the eighteenth century. Estimated at three hundred, their number proves the remarkable expansion of the art, within a few decades, in the French capital.

Julien Le Roy, around 1740, also deplored the leniency shown to candidates for the mastership: "Many Masters unable to make good Watches were admitted; moreover, it has been noticed that when an aspirant sets his sights on being received and feels unable to make a watch, he proposes to the Jurors that he make for his masterpiece an alarm clock with weights; this work, which is nothing in comparison to the repeater that he should make, serves him to be received as a Master. . . . "[148]

So as to calm the always furious fighting between jewelers and watchmakers, royal ordinances specified once again, during the eighteenth century, the privileges and also the obligations of the watchmakers in the realm of watchcase manufacturing. Decrees in 1721, 1722, 1724, 1734 and 1741 reminded the watchmakers that they had to have the title of their works verified in the office of the Jewelers' Guild Hall and that they were obliged to mark the watchcases with their stamp.[149]

THE COURT'S WATCHMAKERS

Besides the master horologists who practiced their trade within the framework of the guild, there were watchmakers to whom the king had granted various privileges. According to Savary des Bruslons,[150] "these watchmakers are not at all subjected to the Jurors' inspections & have, however, the privilege to train apprentices who have a right to reach their mastership & who can be admitted just like other apprentices, the only difference being that they do not pay the fees."

The same historian relates that, among these privileged watchmakers, a distinction was made between those who were "Officers of the King's Bedchamber" and those who had lodgings in the Louvre. The king's officers retained the rank of valets de chambre–horologists. Three in numbers, they had the responsibility of maintaining, regulating and winding the clocks, the pendulum clocks and the watches belonging to the king. To carry out this charge, they accompanied the king to the various houses at which he stayed.

The horologists in the Louvre galleries were generally chosen, in the eighteenth century, for their talents. They enjoyed numerous prerogatives, by virtue of patent letters issued by Henry IV, dated 1608, which allowed them to practice their art under the best possible conditions. Among the king's horologists having lodging in the Louvre during the eighteenth century were Julien Le Roy, Jean-André Lepaute, Alexandre Lefaucheur, François Leloutre, Jean-Antoine Lépine and Robert Robin. The Archives Nationales in Paris possess an interesting file relating to a request from the Parisian guild to limit the privileges of the Louvre clock- and watchmakers, directed particularly against Julien Le Roy, who preferred to occupy his workshop in the rue de Harlay, rather than lodge in the palace.[151]

In this petition addressed to the king on August 20, 1744, the Parisian watchmakers expressed their worry about the rules of apprenticeship, less severe in the Louvre workshops than in those of the city. The complaint against Julien Le Roy—the pretext of the petition against him—was that he had hired a twenty-five-year-old apprentice, Louis-David Carré, for five years on October 22, 1743. The contract went against the regulation according to which no master could accept an apprentice over the age of twenty and also against that which stipulated that the period for apprenticeship had to be eight years.

The masters also worried about the number of apprentices hired by J. Le Roy, on the one hand, in his capacity as "master and elder of the Community" and, on the other, in his capacity of king's horologist. Besides L. Carré, Louis Sénard was doing his apprenticeship at the same time in his shop. J. Le Roy had hired the first one as a master of the Community, the second as a watchmaker in the Louvre. The authors of the complaint asked this question: "It is a matter of deciding whether Mr. Le Roy and those who, like him, will have this double quality, can take three apprentices, one as worker in the Louvre galleries, by virtue of the privileges granted by the patent letters of 1608."

On the margin of the "Observations" made on the request, the following note, dated October 18, 1744, appears: "One should not have any consideration for the watchmakers' remonstrances. However, it would be desirable that no apprentice should be taken above the age of twenty." It was noted that the notoriety of Julien Le Roy had doubtless provoked the envy of the masters of the Corporation: "It appears from the content in their Request that it is Mr. Julien Le Roy, watchmaker, to whom his Majesty granted a lodging in the Galleries . . . who is their sole object in these demands. . . . Apparently, knowing the great reputation this horologist has acquired in the Kingdom and in the foreign countries where his works of horology are diffused, and where even his writings on that art are known, They thought they should avoid directing these demands against him personally, namely, not to appear jealous of his merit. . . . "

THE DIVISION OF LABOR

In the report quoted above, a passage reveals the work organization in Julien Le Roy's shop: "He attracted to him there a very large number of the best workers, whom he keeps busy at the different workings that enter into the composition of the watch. He has them there, within his reach, to guide them in what he asks of them."

This deposition corroborates the information brought forth by Ferdinand Berthoud in the article "Horology" in the *Encyclopédie*, published under Diderot's direction. The function of master is there defined in relation to the activity of his co-workers, watchmakers, journeymen or apprentices: "They are commonly called *horologists*, those who profess *Horology*. But it is pertinent to distinguish the horologist, as it is meant here, from the artist who possesses the principles of the art: they are two absolutely different persons. The first generally practices *Horology* without knowing the first thing about it, & calls himself an horologist because he works on one part of this art. The second, on the contrary, embraces the science in all its scope. He could be called an architect-mechanic; such an artist does not limit himself to one part alone; he draws the plan of the watches and clocks. . . . And as for the execution, he chooses the workers able to execute each part."[152]

The workers employed in the fabrication of watches were called "workers in the small" to differentiate them from those who made pendulum clocks or building and belltower clocks. The *Encyclopédie* article enumerates twenty categories of horology workers specialized in the fabrication of a part of the watch: makers of blank movements, finishers, dialmakers, handmakers, engravers, chasers, gilders, casemakers, enamelers, etc. The artist horologist was supposed to be occupied only in studying and improving the principles of his art, in directing the workers he employed, and in verifying and perfecting the products of his shop.

An important document showing the division of the work in Parisian masters' shops is provided in the sales-book kept by Ferdinand Berthoud and his nephew, Pierre-Louis Berthoud, to whom he gave the management of his shop after 1784.[153] Kept with regularity from 1783 to around 1790, it reveals with great precision the extreme division of the work in watchmaking during the eighteenth century. The various components of each watch sold are indicated, along with the name of the craftsmen who made them. Let us take as an example the repeater watch no. 2070, sold in 1784 to the king's preacher, Abbé Mauri. The wheelwork is by Logeat, the assemblage of the movement by Béliard *fils*, the case by A. Gros, the engraving of the case by Tousin, and the finishing executed by Logeat.

Henri Berthoud died in 1783, leaving his uncle's shop—whose management had been given to him around the year 1770—on the edge of bankruptcy.[154] A notarized act, dated July 14, 1783, tells us that among his creditors were sixteen watchmakers: Jean-Baptiste Alban, Jean-Isaac More, Pierre Béliard, Antoine Demole, Abraham Patron, Jean Jacques Bourquin, Louis Antoine Roggin, Michel Jean Logeat, Daniel Samuel Plattel, Noel Lallemand, Louis Tavernier, Amy Gros, Louis Allier, Jean Amand, Jean Sanguinede, Jacques Vincent Martin.

Master horologists called upon the talents of the craftsmen specialized in the decoration of watches and clocks, of course, cabinet makers, engravers, workers in bronze, enamelers, gilders, etc. The notarized act of Henri Berthoud's creditors also mentions the name of eleven craftsmen co-working with Ferdinand Berthoud's shop and surely with neighboring shops, such as those of Julien Le Roy and of Jean-Antoine Lépine: Edme Roy, caster; André Thil, painter; Alexandre François Mussard, engraver; Jean Baptiste Zaccon, chaser; Jean François Cassin, engraver; Jean François Lorta, sculptor; Jean Tauzin, engraver; Balthazar Liot, gilder; Joseph Cotteau, painter-enameler; Antoine Cochois, bronze modeler; Charles Jacques Tournay, chaser.

THE MANUFACTURES

The workshops of the Court's watchmakers had for their customers the grandees of the kingdom and of the foreign courts. They produced only masterpieces, luxury watches or complicated watches with many indications.

Their production could not satisfy the needs of an expanded clientele. If France had watchmakers of great merit, it was only a tributary of England, and above all of Switzerland, for its ordinary timepieces.

Under the Regency, we saw, the project of establishing a watchmaking manufacture ended in the failure of two attempts, the factory in Versailles and the one in Saint-Germain. It was only in the second half of the eighteenth century that the project was seriously revived.

The two Castel brothers had the idea in 1763 to found a clock and watch factory in Bourg-en-Bresse. They wanted to "challenge the Geneva watch trade, which every year deprives the kingdom of a considerable sum of money and to secure that trade abroad by manufacturing watches at such low prices that the Genevans would not be able to sustain the competition."[155] Watch manufacturing started around 1764, and in August 1766 the king granted the factory the title of "Royal Manufacture of Bourg-en-Bresse." But despite the help, the factory was liquidated in 1777.

A similar undertaking was attempted in Paris. The manufacture established in January 1787 by patent letters of the king, carried the name "Royal Manufacture of Horology." François-Jean Boralle, secretary in attendance to Count d'Artois, was the general manager; watchmaker Jean Romilly was its technical director. Supported by shareholders, the manufacture rapidly lost money. According to E. Gélis,[156] the king went to its aid as early as 1788, granting the sum of 20,000 livres under the condition that it present twenty perfectly trained apprentices. The manufacture was able to continue until 1790. A few private initiatives were more successful than that one during the century.

Voltaire combined a very keen business sense and a profound interest in science. Around 1772, taking advantage of social agitation in Geneva, he welcomed in Ferney exiled craftsmen and founded a watch factory. Lépine, a king's watchmaker, helped with the technical management. All the watch parts were produced in the establishment; four thousand watches, according to Beillard,[157] came out each year without counting the rough sketches of the watch movements to be sent to Paris. After Voltaire's death in 1778, the workers, most of them Geneva natives, returned to their city. The manufacture closed for good in 1792.

The principles of equality and public utility proclaimed by the Revolution served to promote the establishment of watch manufactures that would permit all citizens to acquire inexpensive watches.

A clock and watch factory was created in Besançon in 1793. Its creation was not the result of a government decision, but of a spontaneous gathering of watchmakers and watch traders, revolutionary sympathizers coming from Neuchâtel, Geneva and London.[158] Among the watchmakers there were famous masters such as Auzière, a former chief participant in Voltaire's undertaking, and Lemaire, a specialist in the fabrication of automaton pieces. After surviving with some difficulty on private funds and some aid from the Agricultural Commission, the factory was recognized as being of national interest by a decree passed by the Convention in 1795.

The same decree ordered the foundation in Versailles of a horological establishment where a hundred students were to be instructed each year. The manufacture was not created with the intention of democratizing the art but, on the contrary, with the aim of producing luxury pendulum clocks and watches provided with complications, automata and singing birds. The directors were Constant Lemaire, the project author, a specialist in luxury horology, and Glaesner, a Lyonese watchmaker who had lost all his possessions in the siege of Lyons.

The factory was to be opened in a "National House." After considering the "Garde-Meuble" in the Castle, they chose the house of Madame Elisabeth, the king's sister. In order to repair the damages the house had suffered when it had been used as a military hospital and to install the workshops and lodgings for the workers, important constructions were undertaken on November 20, 1795.[159] Even before the rearrangements started, various measures were taken, aimed at attracting Swiss watchmakers and apprentices to the factory. A decree dated August 16, 1795, promised numerous advantages to the workers: a three-month allocation, starting upon arrival on French soil, of four livres for a bachelor or a widower or six livres for a couple, plus two guineas for each of the first two children and thirty copper coins for the others; the payment of moving costs; and a residential allowance in Versailles. Posters, scattered on the walls in the city, urged parents to send their children, boys and girls, in apprenticeship at the factory.[160]

On April 21, 1796, the plant opened its doors; it numbered thirty Swiss watchmakers and eighteen apprentices. The repairs were not finished. Their costs became so high that the Treasury Commission of the Council of the Five Hundred wanted to suppress the establishment altogether.[161] In August 1796 the home secretary decided its continuance. From then on, the financial situation kept worsening: the manufacture was not receiving any more money; the maintenance and the food of the apprentices were no longer assured; the teachers' salaries were not paid. Lemaire left to settle in Paris; Glaesner lost his mind.

Finally, on March 8, 1801, Bonaparte signed the decree suppressing the Versailles factory. The financial crisis at the end of the century had obviously not helped the undertaking, but other factors had ensured its failure: the very high cost of the repairs and rearrangement of "Elisabeth House" had been harmful from the start to the financial management of the factory; the recourse to Swiss watchmakers—with all the advantages they were given on their arrival—had done much to accentuate the expenses; finally, the organization of production had surely not been as well conceived as in the initial plan, which had claimed to be, before anything, commercial.

The beginnings of mechanized production are linked, in France, to the person of Frédéric Japy (1749–1812), thanks to whom the Montbéliard region, French since 1793, went through a period of prosperity founded upon the production of movements in the rough. In 1770 Japy started a shop in Beaucourt to rough-cast mechanical parts, where the various machines he had invented for fabricating pilars, plates, spring barrels, gears and balance-wheels were used. Forty thousand roughcasts were leaving the factory in 1795. Another plant was created in 1793, at Fontainemelon.

French horology and the other horological centers in Europe

Renown of the French Horologists

The diverse initiatives that helped develop the watch trade gave international renown to French horology. Despite the strong competition offered by the Swiss and the English, French masters were admired throughout Europe for their science and skill.

In a brochure published in the Hague in 1767, titled "Reflexions sur l'horlogerie en général . . . ," a king's watchmaker, Béliard, summed up in these words the reputation of French horology: "The perfection to which, over the past forty years, french Horology has been brought; the reputation it has acquired Abroad has made it the object of a very extensive trade. The English, whose industry was formerly so superior to ours, & who furnished us watches for which considerable sums of money left the Kingdom, not only do not send us any anymore, but import many from France. Finally, French Horology has acquired such a reputation in the whole of Europe, that most Nations have hardly any watches other than French ones, or imitations of them."

The article "Horology" in the *Encyclopédie* gives us interesting evidence that, while asserting the supremacy of French watchmaking, exposes the problems and the uncertainties of its situation: "Though Horology has now been brought to a very high degree of perfection, its position remains critical; for if, solely through the love of a few artists, it has reached a degree of perfection much above that of the English, on the other hand, it is ready to fall back into oblivion. . . . [T]he trade done by the merchants, workers without rights or talents, domestics & other intriguers who deceive the public with false names, which debase this art; all these things gradually erode the confidence that one had in the famous artists, who, finally discouraged & carried away by the torrent, will be obliged to do like everybody else, cease being artists and become merchants."[162]

Many watchmakers were expressing this fear of seeing their art depreciate because of the trade of haberdashers or jewelers. These attracted their customers by the ostentatious display of their goods, which more often than not were made up of defective watches and pendulum clocks, hastily fabricated in Swiss manufactures or made illegally by apprentices unable to accede to mastership. These pieces were sometimes signed with the most illustrious names. Julien Le Roy and Jean-André Lepaute would complain of these merchants, who used their names in watches whose movements were neglected for the sake of carefully decorated cases. These falsifications ran the risk of compromising not only the internal watch trade but also the export of French watches. Watches falsely signed Julien Le Roy were directed, according to Tardy, largely toward the American colonists, who ended up by refusing them.

In order to restore to the art the confidence it deserved, the creation of a "Horological Society or Academy" was put forth in the article "Horology" in the *Encyclopédie*[163]: "Then the public, being instructed about those they should trust, would cease buying horology works from deceiving merchants, assured to find at the artist-craftsmen's shop only excellent machines; finally, as a result of these various advantages, the perfection to which our horology has been brought would be so well known that the foreigner would altogether prefer it to our neighbors'".[164]

The high repute of French horologists spread throughout Europe. Their watches found enthusiastic customers in every foreign court because of the movements' perfection and the refined decoration of the cases. The splendid diamond watch matched with a chatelaine, made in 1767 for Queen Carolina-Mathilda of Denmark, is signed by a French watchmaker; the decoration is the work of the French jeweler Jean-François Fistaine brought to Copenhagen by Christian VII.[165]

French watchmakers went to settle abroad, sometimes answering a royal solicitation. Joseph Gay became horologist to King Vittorio-Amedeo III, in Turin at the end of the century; he made the watch on the bottom of which the portrait of that king is painted on enamel.[166] Fol, third valet of the King from 1777 to 1784, became horologist to the king of Poland, Stanislaus-Augustus, in 1786.

On the eve of the Revolution, General George Washington attested in a letter to his friend Gouverneur Morris, the qualities then recognized in French watches: "—I am much obliged by your offer of executing commissions for me in Europe, and shall take the liberty of charging you with one only.—I wish to have a gold watch for my own use (not a small trifling nor finically ornamented one) but a watch well executed, and about the size and kind of that which was procured by Mr Jefferson for Mr Madison (which was large and flat).—I imagine Mr Jefferson can give you the best advice on the subject, as I am told, these species of watches, which I have described, can be found cheaper and better made in Paris than in London." So as to accomplish his mission, Morris went to the place des Victoires, to Jean-Antoine Lépine who sold him two gold watches of which one was handed to the general through the offices of Thomas Jefferson.[167]

Foreign customers appreciated above all the French decoration of horological pieces. A passage of the "Horology" article of the *Encyclopédie* states: "We possess to the utmost degree the art of tastefully ornamenting with taste our pendulum watches and watch cases, whose decoration of which is much superior to foreigners'."

THE SWISS COMPETITION

If only they had been satisfactorily run, the clock and watch manufactures would surely have served to efficaciously combat the Swiss competition, which, according to contemporary evidence, was a true scourge. Towards the midcentury, Julien Le Roy noted: "The Genevans are flooding all the southern provinces of the kingdom with their bad watches. Besides the Merchants who carry them for sizable sums to the Beaucaire Fair, there are many others who each summer carry large quantities of them in Dauphiny, Provence, Languedoc and Gascony. While it appears difficult to remedy those inconveniences There, one could at least Try, by establishing two factories, one near Bordeaux or Bayonne, to supply those quarters and maybe Spain in the future, and the other in Provence as much to supply it as with the view of supplying the Levant and Italy."[168]

The vigor of the Swiss competition was mostly felt in Paris, where the Genevan *établisseurs* had numerous customers among the merchant jewelers and the watchmakers desirous of increasing their sales with cheap prices. The public was deceived about the quality of the Genevan watches in several ways: (1) by the signature, which often borrowed the name of one of the best masters in London or Paris; (2) by the case, whose standard was inferior to the legal French gold standard; (3) by the generally defective movement; (4) by a lower price than those of French watches, which was indeed the result of poorer silver or gold standards and hasty execution of the movement.

Julien Le Roy stressed the importance of the *établisseurs'* role and described thus their contrivances: "These are the watch merchants who give employment to most of the Workers in Geneva, whom they always recommend to pay attention more to the beauty of the Cases than to the perfection of the movement: by this means they succeed very well in making the public fall into a trap that they hide under the deceptive outside appearance of beauty and a bargain price, while most of their Watches could not even run three months without breaking down."[169]

In France the gold standard was fixed at twenty carats, while in Switzerland it could be arbitrarily determined by the craftsman. The leniency of the Swiss law promoted an important traffic of watchcases between Geneva and Paris. A legal memorandum, kept in the Bibliothèque Nationale in Paris, explains the functioning of this illegal traffic.[170] In the best instances, the cap of the case was in the French standard and carried the regulation hallmark of the Common House office, but the other, smaller parts, such as the rim, the hinge, the bezel and the pendant, exempted from receiving the hallmark, were manufactured in Switzerland according to an inferior standard. Let us refer back to the author of the memorandum: "We saw that in the Watchcase made either of silver or gold, only the *Cap* can be countermarked, after testing, with the hallmark of Paris. We saw that this testing is done on a Plate wrought for the purpose and still in the rough. Finally, we saw that the *Set Parts*, which are the components of the case, weigh much more than the Cap. What do the Foreigners do, who meddle in that trade? They send Plates to Geneva countermarked with the Paris hallmark to fabricate Cases with them. These Cases hence are done in Geneva, only the *Cap* is in the standard of France. The rest, that is to say the Bezel, the Rim, the Hinge, the Pendants & other set parts, has no determined standard because in Geneva, each Worker works with the standard he pleases. . . ." On their return to France, the watchcases thus fabricated could be sold cheaper and in complete legality, since they carried the French hallmark certifying the standard.

Often, the lure of easy money enticed some into other dishonest practices. According to the author of the memorandum, "the Cases come back to France provided with as defective movements as are the standards of gold and silver. . . . When these Watches with gold or silver cases arrive in Paris & reach the hands of the Masters who meddle with that Trade, they engrave their names on the plate of the movement. Thus, since one sees on the case the stamp of the hallmark & on the movement the signature of a Paris master, these watches pass as being made in Paris, & their delivery is so much easier that individuals fancy themselves to be the less deceived: besides, bargain prices always facilitate the sale. . . ."

In the watchcases already mentioned, the cap was of the French standard. But there also existed entirely Swiss-made cases. The Genevan *établisseurs* managed to sell them in France using a punchmark enforced by law since 1679. Called *poinçon de reconnaissance*, it had been established to mark foreign plates and silverware; in no way did it certify the standard of the piece. Theoretically, it should not have been applied to watchcases, since all had to bear the stamp certifying the regulation standard. But the warden-goldsmiths, on the payment of a fee, engraved this mark on cases of foreign origin they were presented; despite the inferior standard.

In his *Réflexions* (op. cit.) Béliard showed to what extent this illicit trade could harm French watchmaking: "The unlimited and unrestrained introduction of so many watches from Geneva and Switzerland, has harmed in all kinds of ways the horology of France; the watches sold ready-made, by the mediocrity of their work, the lightness of their cases and the substandard gold, have accustomed the public to so base a price that most watchmakers, in order not to lose too greatly in trade competition and in order to be able to sell their watches, have seen themselves forced to lower their prices and consequently diminish the quality of their work, or even to use some of these—not recognized—Watch Movements from Geneva."

The Swiss competition never ceased to increase its power. Interesting statistics showing that growth are given in a *Notice sur l'Horlogerie de Paris*, published in 1840 by Charles-Louis Le Roy, the king's horologist. In 1786 the sales of his grandfather, Basile-Charles Le Roy, which amounted to about three hundred per year, were as follows: fifty watches were entirely made in Paris; one hundred others used Swiss movements, the cases, dials and winding mechanisms being from Paris; the last hundred and fifty were entirely made in Switzerland. The proportion of Swiss watches increased strongly in the sales of the first quarter of the nineteenth century; according to Le Roy, "[F]rom 1800 to 1814, out of 500 to 600 watches sold each year [by his father], 30 at most were entirely made in Paris."

The flowering of the horological art in Switzerland

Prosperity of the Genevan Factory

Horology in Geneva went in the course of the eighteenth century through an expansion that was to last well beyond that period. Around the midcentury, the number of watchmakers in Geneva, was estimated at seven hundred.[171] Forty years later, they were more than a thousand.

The "Ordinances for the Art of Horology" were revised by the Council of the Two Hundred on July 13, 1731, and then on September 11, 1745.[172] The minimum duration for apprenticeship was set at five years. The journeyman who wanted to become a master had to address the "Lord Clerks and the Jurors, so that upon his request, they make the Corps of Masters assemble to order from him a Masterpiece that shall be an Alarm Clock or a Repeater." Several articles reminded with authority that the watchmaking trade was forbidden to women. It was prohibited "to all masters, Companions & others to instruct their wives or daughters to do Horology work . . . it being only permitted to make the Open-Work, the Hands, the Pilars, the Chains, the Spirals, the Keys, the Polishings & to cut again the Wheels & the Fusees & be able to gilt the Watches." The division of the work became more pronounced during the eighteenth century. At the end of the century about thirty trades concurred to the production of watches and clocks. Without counting the spring- and the chainmakers, which had appeared in the seventeenth century, watchmakers were assisted by makers of pinions, verges and striking systems. Goldsmiths and silversmiths, case assemblers, engravers, machine turners, jewelers, gem cutters and enamelers were employed in the fabrication and the decoration of watchcases. To these craftsmen one had to add the dialmakers, the makers of hands, the polishers, gilders and finally the *établisseurs*, who coordinated the various activities of the Factory and took charge of the sales.

The first rigorously conducted census of the Factory dates from 1788. It was published in a supplement to the *Journal de Genève*, on March 14, 1789.[173] The Watchmakers' Guild was composed of 1,095 members; the Engravers, 204; the Case Assemblers, 475. During the eighteenth century, no other guild was created, though the number of active craftsmen, in certain branches, could have justified the appearance of a new community. According to the census, the Factory included 78 jewelers, 72 enamelers, 113 springmakers and 31 merchant horologists. It is estimated that it must have been employing three thousand people at least.

Until the start of the nineteenth century, production was not organized in an industrial manner. Small workshops called *cabinets d'horlogerie* grouped a few watchworkers, the *cabinotiers*, under the direction of a master. The workday was between ten and twelve hours. This type of production, apparently archaic but modernized through the distribution and coordination of tasks by the *établisseurs*, greatly ensured the success of Genevan watchmaking.

Watchmaking begins in the canton of Neuchâtel

During the whole second half of the eighteenth century and at the start of the next, watchmaking in Neuchâtel went through a strong expansion, as attested by a few statistics given by Chapuis[174]: in 1752 there were 464 Neuchâtelese watchmakers; in 1764 they were 1,023; in 1788 their number rose to 3,634; in 1808 it reached 4,316. This rapid expansion is explained by various causes often invoked: the barrenness of the soil, which obliged agricultists to have a second activity; the length and the roughness of the winters, which impelled the inhabitants to work at home; the cheap price of the watches thus produced.

In Neuchâtel, watchmaking was a fairly prosperous activity in the first half of the eighteenth century. Its development had been favored by the coming, after the revocation of the Edict of Nantes, of Huguenot refugees, such as the Parisian Jacques Dupuy. The most eminent watchmaker of that period in Neuchâtel was Josué Sibelin, received as a master in 1689.

The severity of the Guild's regulations, which admitted only the bourgeoisie among its members, probably contributed to the decline of watchmaking in that city, despite the creation of a manufacture not only to produce pieces of horology but also to train watchmakers. Managed by an excellent master, D. Jeanrichard, the factory was no more fortunate than establishments of its kind in France.

The beginnings of small-size horology in the mountains surrounding Neuchâtel are closely associated with Daniel Jeanrichard's activity. This jeweler-watchmaker had the idea of organizing "in the mountains" the *établissage* as it had been practiced in Geneva. He put together a real factory, made of several workshops whose activities complemented each other, and took the responsibility of selling the pieces they produced. When D. Jeanrichard died in 1741, hundreds of watchmakers were working in the valleys of the Locle and of La Chaux-de-Fonds.[175]

D. IEAN-RICHARD
AV LOCLE

25 Watch movement signed "D. Jean-Richard." The Locle, early seventeenth century. D. 44; Th. 19.
Daniel Jeanrichard (1665–1741), established in Le Locle around 1700. The beginnings of watchmaking in Neuchâtel are closely linked with his activity both as a watchmaker and as an *établisseur*.
La Chaux-de-Fonds, Musée International d'Horlogerie, inv. 25

Famous *établissages* were started in the second half of the eighteenth century. The House Philippe Dubois and Son, established near the Locle, began a rapid expansion in the last third of the century; it gave work to several hundred craftsmen, among whom were sons of Daniel Jeanrichard.[176] In La Chaux-de-Fonds, the Courvoisier House, long managed by Louis Courvoisier (1758–1832), had an important commercial activity at the end of the eighteenth and the first half of the nineteenth centuries.

By the end of the eighteenth century, watch- and clockmaking were already solidly implanted in the small towns of the Val-de-Travers. There were a dozen important workshops in Fleurier, Travers and Motiers.[177] In the first half of the nineteenth century, the industry went through a new expansion, caused by the manufacture of watches for the China trade; Fleurier became a real center, specializing in the production of "Chinese watches."

The watchmaking industry continued to expand in Swiss territory and flowered, in the first half of the nineteenth century in Tramelan, Bienne and Sainte-Croix.

Commercial dynamism of the merchant-watchmakers

While production increased and the division of work was spreading, the *établisseurs*' role became more and more important in management and commercialization. Several ordinances regulated their activities: they had to keep an exact register of all their negotiations; their office lasted three years but was renewable.[178]

The *établisseurs* from Geneva, the Locle and La Chaux-de-Fonds assiduously frequented the fairs of Beaucaire, Lyons, Frankfurt, Basle and Utrecht. They stepped up the number of commercial contacts with the traders of the main cities in Europe. Their clientele in the countries neighboring Switzerland was very important.

We have already had the opportunity of stressing the importance of the number of Swiss watches sold in France. Commercial relations with England grew during the second half of the eighteenth century; Swiss watchmakers and *établisseurs* settled in England in order to control and increase their trade. London watchmakers, like their Parisian colleagues, ordered movements and cases from Swiss merchants. Holland and Belgium offered important outlets to the Swiss industry; many watches were exported to Amsterdam, either to augment the trade of the city's horologists or to be sent to faraway colonies, such as Java.

If the *établisseurs* had found the way to corner the watch market in countries with flourishing horological industry, what could they not do in countries that did not have any watchmaking? In Germany, where small-work horology had lost its former splendor, the Genevan and Neuchatelese productions were dividing among themselves the market of the Southern States, of the Rhine courts as well as Berlin. The Swiss *établisseurs* took root with equal success in Italy and Spain.

Under the auspices of Catherine II, Swiss traders started the conquest of the Russian market. The trade with Russia increased in the first quarter of the eighteenth century, thanks to the activities of the House of Robert & Courvoisier in La Chaux-de-Fonds and the House of Paul Buhre of the Locle. The latter established a store in Saint Petersburg that acquired such great renown that Paul Buhre received the privilege of being the "official purveyor of Her Majesty."[179] The Swiss *établisseurs* succeeded also in planting themselves in Northern Europe, notably in Denmark.

Trading in Swiss watches in the Near East developed during the second half of the eighteenth century. Numerous watches were distributed throughout the Turkish Empire from Constantinople and Smyrna, which played the role of agents. In that trade, the Swiss met English competition. But here, as elsewhere, they were able to take advantage of the situation: many watches sold to the Turks by the London merchants had been, in fact, made in Geneva.

At the end of the eighteenth century, China was already an important market of Swiss production.[180] The traders had been, on the emperor's order, regrouped in Canton; they could trade only with Chinese merchants called *hannists*. Charles de Constant de Rebecque (1762–1835), a cousin of Benjamin Constant, was among these courageous men. The Library of Geneva has kept most of his as yet unpublished writings, where he underlined the difficulties met by European businessmen in the factories of Canton. The chief Customs officer, the "hoopoo," abused his commission by imposing excessive entry taxes and receiving many bribes. The *hannist* merchants, entrusted by him to register goods before unloading, would commit all kinds of dishonesties at the expense of the Europeans: thefts and false estimations of the value of goods. Despite all these difficulties, the Swiss watch trade in China developed considerably.

The workshops in Fleurier and throughout the Val-de-Travers specialized in the production of watches for the Chinese trade. The initiator of that activity was Edouard Bovet (1797–1849). In January 1820 this trader, a native of Fleurier, formed a society "aiming at the horological trade with China."[181] The youngest of the Bovet brothers, Charles-Henri, born in 1802, was received in the partnership in 1824. The House, established in Canton, was directed by Edouard and Charles, then by Charles alone, from 1830 to 1836, when he was joined by his nephew, Louis, who was to take the management of the business.

The success of the Bovet House is thus stated by Chapuis:[182] "For a long time among the Celestials, a watch was a Bovet, and that Bovet watch would be used, even way inside the country, as 'money' in the exchanges."[182] The example of the Bovets was followed by many *établisseurs*, such as the Vaucher brothers, the Dimier brothers and the Juvets.

Beginnings of mechanization

The success of Swiss watchmaking rested largely on the extreme division of labor organized by the *établisseurs*. This distribution of tasks had promoted the appearance of workshops specialized in the manufacture of movements in the rough. One measures the importance of such workshops by reporting that in France their absence was one of the fundamental causes that prevented watchmaking from turning toward mass production and made it a tributary of Swiss industry.

In the first half of the nineteenth century, Swiss watchmakers considered the manufacturing of *ébauches* (rough drafts) that would allow them to increase production and lower the price of their products.[183] This mechanization was not imposed without opposition. Thus, in Geneva the watchmakers showed considerable reticence toward mechanization. They did not want their *cabinets* replaced by large workshops that would deprive them of their freedom. Various attempts at the end of the eighteenth century to create a manufacture in Geneva failed. The first *ébauches* factory that succeeded in asserting itself was the one of Sandoz, in partnership first with Trot, then with Rossel, created in 1804.[184]

The vigor of English horology is maintained

This period, which corresponded to the conquest of precision and the appearance of the modern watch, was again a fortunate one for English watchmaking. In the third quarter of the eighteenth century, the Clockmakers' Company of London received approval for divers measures that helped its expansion. Greatly talented masters pursued the researches undertaken by their illustrious predecessors. The flourishing state of English horology was above all manifested by the prosperity of national trade, which had spread from the Near East to the Far East.

Measures favoring the Clockmakers' Company

Around the mid-eighteenth century the Company became seriously worried by the diminution of its membership. The directing committee decided then to raise a campaign of petitions addressed to the lord mayor and to the Court of Aldermen, so as to obtain their approval for measures that would remedy the ills of the Company.

For several dozen years the Court of Assistants had wished that any person practicing the art be obliged to belong to the Clockmakers' Company. Such a decision could indeed favor an increase in the number of watchmakers. The Assistants' demand was finally accepted on December 1765. The act of consent stated: "[I]t is enacted, ordained and established by the Right Honorable the Lord Mayor of this City of London . . . that from and after the Twenty-fifth day of December 1765 every person not being already free of this City using or exercising or who shall use or exercise the Art Trade or Mystery of Clockmaking within this City of London or liberties thereof shall take up his or her Freedom and be made free of the said Company. . . ."[185]

According to the Court of Assistants, the main reason of the decreasing adherences to the Company rested on the fact that the Clockmakers' Company had not obtained the effective creation of a livery company, as was the usage in other guilds or communities in the city. The statutes of the Company, approved in 1632, indeed included that clause, but it had not yet been ratified by the Court of Aldermen. On January 31, 1748, the master, the wardens and the assistants presented a petition in these words: "That it having been found by experience that the not being of the Livery hath been in many instances of great disadvantage: – that the number to be bound Apprentices and others to be made free either by servitude or purchase is considerably decreased which your Petitioners cannot ascribe to any other cause than that before mentioned. Your Petitioners therefore pray your Lordship and this Honorable Court to grant them the honor and the priviledge [*sic*] of the Livery of this City in order to remove those inconveniences. . . ."[186] The petitions succeeded each other in vain until 1766, the year when the Court of Aldermen granted the creation, among the Company, of a livery whose members could take part in all official ceremonies in the city. The number of masters would not be over sixty. This number was increased to one hundred twenty in November 1786, then to two hundred in 1810.

Talented master craftsmen

All the worthiest representatives of English watchmaking in the second half of the eighteenth century took part in the major discovery of that period: the perfecting of a timepiece whose precision would permit the determination of longitudes at sea. Encouraged by Graham, John Harrison (1693–1776) started researches on marine chronometers. His models were followed by those of Thomas Mudge (1715–1794), then by those of John Arnold (1736–1799) and of Thomas Earnshaw (1749–1829), who gave the marine chronometer its modern form.[187]

These renowned masters worked for all the courts of Europe, and above all for the one of George III.[188] A number of the luxury watches executed for the king of England became famous for the technical novelties or for the complications they offered. These watches were, in many ways, unique. Thus, Thomas Mudge made the first watch with an anchor escapement. In 1764 John Arnold presented his king with a tiny ring watch with a repeater and a cylinder escapement made of ruby. Though he was offered 1,000 guineas by Empress Catherine II for a similar one, Arnold refused the order.

Thomas Mudge acquired great fame for watches provided with complications. The king of Spain, Ferdinand VI, ordered him to build a very complicated watch, giving the horologist the freedom to choose the complications that would interest him the most. Mudge sent the king a watch whose movement was lodged in the pommel of a cane. It indicated both real and apparent time, struck the hours and had a repeater striking hours, quarter-hours and minutes.

In the French court, English watches were still very much appreciated. We discovered interesting evidence of that fact in the Archives Nationales.[189] In December 1786 Louis XVI ordered the payment of 2,778 livres for a watch he had ordered from Josiah Emery (1725?–1797) through his own watchmaker, Robin. The watch had been delivered on August 30, 1786, to M. de Saron, who informed the count of Angivillers: "Mr. de Saron is honored to let Monsieur d'Angivillers know that he received today the gold watch or chronometer of S[ieu]r Josiah Emery. Mr. de Saron received it only at one o'clock in the afternoon and believes to be already in the situation to warn Monsieur d'Angivillers that it is five to ten seconds early on mean time every twenty-four hours, which agrees with what the English watchmaker let him know. He regulates them so, at the time of delivery, because he observed that after one year, they become late for a time about equal to that quantity and afterward run constantly well. . . ." Among other observations, we pick up this one: "Mr. de Saron fears that it will be found a bit heavy; it weighs close to half a pound. On this point the English do not think to economize on the material."

The excellence of English horology was not only manifest through creations of a technically superior level but also through the many improvements brought to pendulum clocks and ordinary watches. The care for quality constantly appears in the pieces made in the eighteenth century by great English horologists, such as John Ellicott (1706–1772), watchmaker to George III, who contributed in developing the use of the cylinder escapement, and Josiah Emery, the author of numerous watches with an anchor escapement.

The English horological trade

Through the eighteenth century and at the beginning of the nineteenth, English horology kept the international renown it had acquired during the years 1680 to 1730, thanks to masters like Tompion and Graham.

The export of English watches in France was still active around the mid-eighteenth century, as is testified by Julien Le Roy: "Excessive [customs] rights favor the English, whose Ship Captains bring Watches to every port in Brittany, doing this all the more handily since Watches are easy to hide. . . ."[191] The trade was nevertheless reduced by the increase of French production and by the activity of Swiss *établisseurs*, who undersold everybody. The English did not do any better than the French in organizing the production of ordinary watches the middle classes could afford.

The English watch trade developed considerably in the Near East in the second half of the century. The sales were particularly important in the Turkish Empire. J. Dallaway stated in 1797: "English watches, prepared for the Levant market, are more in demand than those of other Frank nations, and are one of the first articles of luxury that a Turk purchases or changes if he has money to spare."[191]

Many English watchmakers specialized in the production of watches meant for the Turkish trade. Some drew important profits from that commerce by adding to their export activities those of Eastern imports. George Prior and Isaac Rogers, who exported thousands of watches to Turkey, received in return, exotic products.

Robert Markham managed a workshop producing numerous watches for Turkey. The number of these watches bearing his signature shows the importance of his company. Markham added to his name the one of his father-in-law, James Markwick, who had given him the management of the business. Other names at

26

27

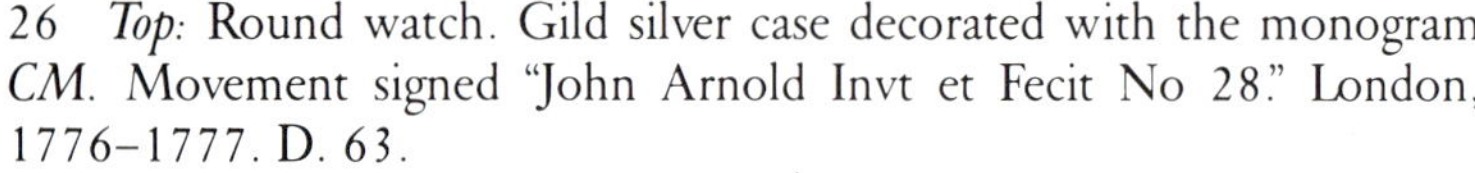

26 *Top:* Round watch. Gild silver case decorated with the monogram *CM*. Movement signed "John Arnold Invt et Fecit No 28." London, 1776–1777. D. 63.
This watch made by John Arnold (1736–1799) is the oldest known having a pivoted detent escapement. The *S*-shaped balance is a slightly posterior modification.
Bottom: Movement signed "Thos Mudge London 260." 1755. Rear of the case. D. 48.
Thomas Mudge made the movement for a watch sold, on August 22, 1755, to the king of Spain. It has two spiral springs; the bimetallic device for thermal compensation acts on one, the regulating rack on the other. An equalizing remontoir ensures the transmitting of the motive power to the escapement.
London, Guildhall library, The Clock Room, inv. 419 and 189

27 Round watch with a pair case. Movement signed "Markwick Markham Perigal London" and numbered 16656. London, third quarter of the eighteenth century. D (exterior case) 38; Th. 15.
The white enamel dial has Turkish numerals. The striped agate exterior case is covered by an engraved gold lattice.
Robert Markham was the manager of an important shop, specializing in the manufacture of watches for the Turkish trade. He added to his name the one of his father-in-law, James Markwick, who had left him the business. Perigal is the name of one of his partners.
Paris, Louvre, Olivier Coll., inv. OA 8546

28 Peach-shaped "Chinese" watch. Painted enamel gold case. Movement signed by Ilbery and numbered 6498. London, first quarter of the nineteenth century. L. 60; W. 53. ▷
William Ilbery (fl. 1780–1839) became famous in the manufacture of highly refined watches meant for the Chinese market.
Rockford (Illinois), Time Museum, inv. B 179

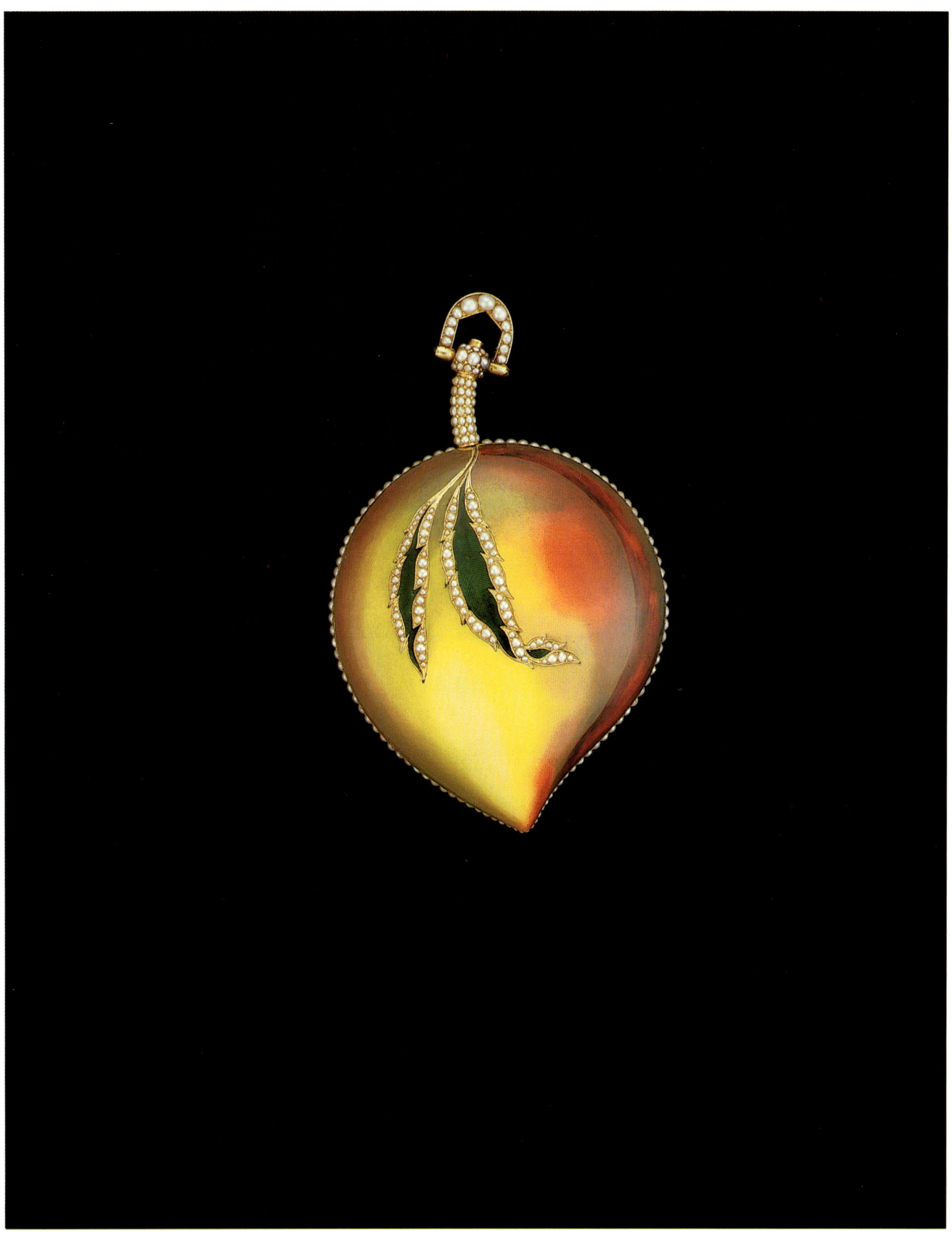

times complete this double signature; one can meet in particular watches signed "Markwick Markham Perigal", "Markwick Markham Borell", "Markwick Markham Recordon." The partners, for an undetermined time, of R. Markham may have kept the right, after the death of the master, to use the name "Markwick Markham," whose reputation in Turkey secured them a prosperous trade.

Thanks to the East India Company, the English developed their trade with China during the eighteenth century. Sales of watches and pendulum clocks to Chinese officials were very important. They took place through the intermediary of British traders established in Canton. The trade was difficult and even dangerous. Charles de Constant relates this story: "An English supercargo had several very beautiful pendulum clocks the mandarins desired, which led him to ask for a rather stiff price. The merchants with whom we do the trade offered him half only of what he asked. He refused to sell for that price; consequently, he received the order from the mandarins to leave the country and take the pendulum clocks back with him, with no one permitted to buy them, which forced him to let them have them at their offered price. . . ."[192]

The profits that could be gained through the China watch trade moved some London watchmakers to specialize in the traffic as early as the mid-eighteenth century. James Cox founded a famous store in Canton, offering rich customers luxuriously decorated watches, pendulum clocks and automata. At the start of the nineteenth century, William Anthony (1764–1844) and William Ilbery (1780–1851) successfully specialized in the manufacture of watches meant for China. Ilbery's production is famous for its splendidly decorated paintings on enamel. W. Ilbery was without a doubt the creator of the "Chinese watch," primarily characterized by a movement richly engraved with foliages and interlacings.

Established in London before moving their workshop to Fleurier, the Bovets were the first imitators of Ilbery. If, like Ilbery, they created luxury watches, they above all produced good-quality watches with an approachable price for the Chinese middle class. Here again can be seen the trading spirit that made Swiss horology succeed at the expense of the English and the French.

V From 1830 to 1900: Transformations in the watchmaking industry

As industrialization developed, watches became indispensable objects in daily life. The appearance of railroads, the rigorous organization of work schedules and the multiplication of commercial exchanges created a new way of life that necessitated the ownership of a watch. In order to respond to the growing consumers' demands, watchmakers had to change their work methods.

The visitors in the 1889 World Exhibition were able to note the "complete transformation of work methods" in the domain of horology.[193] On the occasion of the Exhibition, a Chronometry International Congress took place, in the course of which Paul Garnier, horologist for the French navy and the railroads, presented a report on the changes happening in watch manufacturing. Of note is his statement: "The manufacturing by machines, or mechanical production, is being applied not only on cheap watches but on the precision watches. The perfection of the parts is such as to allow their interchangeability. . . . "[194]

Industrial manufacturing was applied successfully above all in the production of ordinary watches. In 1889 the most important plant specializing in this area was the one of "Japy frères et Cie" in Beaucourt and Badevel, which yearly produced three to four hundred thousand watches with a key or a winder.[195]

The new work methods were rapidly made use of by purveyors of horological parts such as springs, dials, hands, glasses and spirals. The 1889 Exhibition numbered fifty-five French exhibitors and thirty-six Swiss who were presenting good-quality parts obtained through the new mechanical processes.[196]

Such changes were partly promoted by the increasing production of newly created factories in the United States, which threatened to reduce the outlets of European horology.[197] A witness analyzed the situation thus: " . . . the Americans, wanting to escape the yoke of European monopoly and lacking a sufficient number of good workers, attempted to ask the machine what only the hand had provided until then. Gigantic factories were mounted. . . . [A]fter long efforts, these factories were able to deliver good watches to commerce. . . . The watch manufacturers of Switzerland, seeing important markets shut in front of them, doubled their efforts, and since skillful workers were numerous among

29 *Ebauches* of watches made by the Japy factory in Besançon (Jura) nineteenth century.
Paris, Musée National des Techniques (CNAM), inv. 8367

them, they decided to introduce machines anywhere they offered a serious advantage."[198]

The decision was followed by fortunate results: Switzerland acquired, through the nineteenth century, a dominant position in watch manufacturing. Its most important center was Geneva, where, in 1869, seven thousand persons were employed in the watchmaking industry. The city's yearly production was at least as high as a hundred thousand watches, of which eleven out of twelve were made in gold; to this should be added the thousands of movements meant for export.[199] The towns of the Locle, of Fleurier, of La Chaux-de-Fonds also had very active workshops, producing pocket chronometers, precision watches and ordinary watches. In 1866 in the canton of Neuchâtel, 13,706 workers produced about seven hundred thousand watches per year.[200] In the canton of Vaud, the Joux Valley was famous for the *ébauches* from which watches with complications were made.[201] The canton of Berne, whose watchmaking center was Bienne, numbered thirteen thousand persons linked to a watchmaking industry particularly oriented toward the production of high-quality but inexpensive watches.[202]

During the nineteenth century, France kept an important place in the manufacturing of watches, developed particularly in Besançon. The factory, there, was specialized in the mounting of watches from *ébauches* bought outside, notably from plants in the Montbeliard region specializing in the purveyance of these parts.

A *Notice sur la fabrication de l'horlogerie à Besançon* was published in 1867, on the occasion of the World Exposition. The writer made these observations: "The watchmaking population in Besançon is estimated at nearly 15,000 souls. . . . The horologic work is divided into an infinity of segments, each of which forms a small shop made up most often of the members of a family. . . . Each of these shops is composed of 7 or 8 workers. . . ." Such organization, which no longer corresponded to market necessity, harmed the development of watchmaking industry in France. It went into an irremediable decline around 1900, to the advantage of its Swiss competitor, who could foresee new successes in the twentieth century.

Part two

Technical evolution

I The primitive watch movement

Originally, the watch movement was contained in a frame formed by two parallel plates separated by pillars. It was then made of iron and was composed, like the table clocks', of a spring, wheels, an escapement with a crown wheel, and a straight foliot. Around the mid-sixteenth century, various improvements were introduced: the use of brass instead of iron, generalization of the fusee to regularize the motive power, adoption of an annulary foliot and the use of screws instead of pins. Since then, no other essential part has been added.

Description of the main components

A hand-forged spring provides the motive power of the movement. The metal intended for making it was hammered, drawn, and then filed to give it an even thickness.

The driving spring is lodged in a cylindrical box, named for its shape the *barrel*. An axle runs through the barrel. An "arbor" is formed, at its center, by a cylinder that fills about one third of the inside volume of the barrel. The spring occupies half the remaining space; one of its extremities is fixed to the arbor, while the other one, ending with a hook (called the "English hook" by French and Swiss watchmakers in the eighteenth century) is hooked to the ferrule of the barrel. When the movement is wound up, the spring tightens closer against the arbor and leaves a space between its last coil and the ferrule. The motive power that activates the movement is produced by the straining of the spring, which, through its elasticity, tends to reoccupy its space, causing the barrel to turn round. The pointed end of a small lever called the *pawl* is engaged between the teeth, sharp and at an angle, of a ratchet wheel mounted on top of the arbor of the barrel. The pawl maintains the tension of the spring by preventing any backward motion. Toward the end of the eighteenth century, the pawl and ratchet-wheel system was replaced by the *worm* and the wormwheel.

At the start of the seventeenth century, watches ran an average of about sixteen hours on a windup. Abbé Develle was able to

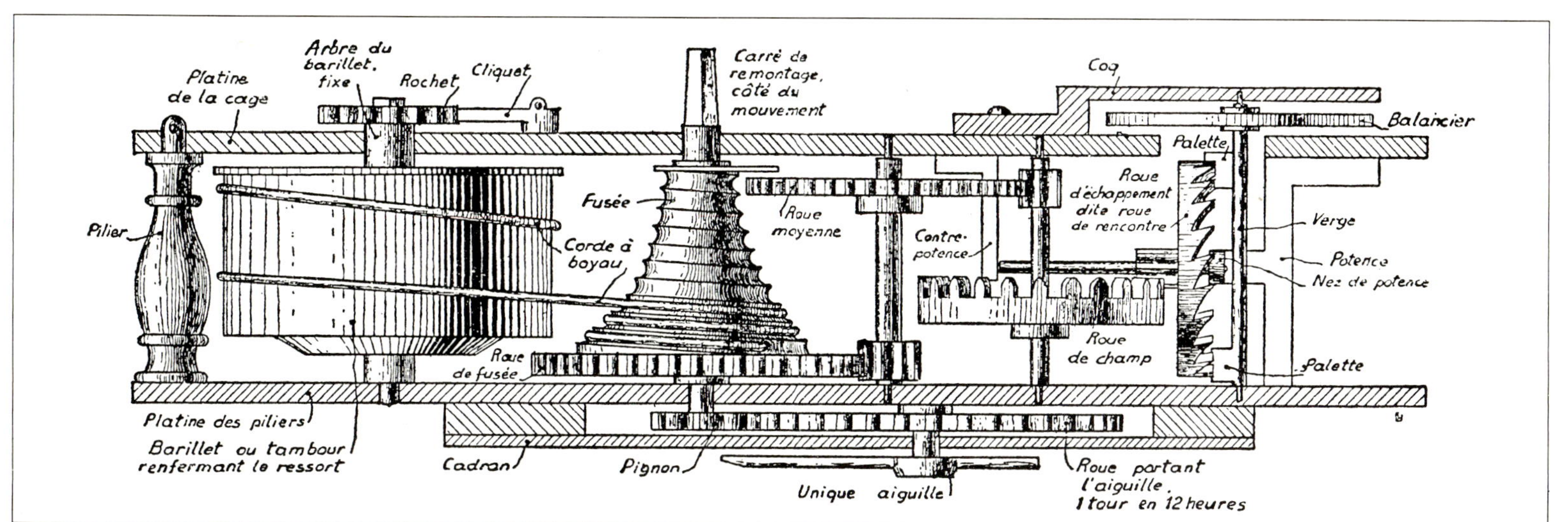

30 Primitive watch movement. The foliot is protected by a cock equipped with one foot. The fusee has a gut cord. The dial has only one hand, indicating the hours.

examine many contracts in which the duration of running time wished for by the customer is mentioned: out of nineteen such contracts, dated from 1605 to 1626, fourteen require a watch going from fifteen to sixteen hours, four desire twenty-six hours, and only one demands a watch going thirty hours.[1]

These watches ran with very little precision. A watch was about thirty minutes fast during the few hours after winding and was equally slow when the driving spring was almost unwound. The necessity for frequent adjustments explains the presence of a small sundial inside the lid of numerous watchcases.

The payment for a delivery made by the horologist Abraham de La Garde to King Henry III of Navarre gives evidence of the vogue at the end of the sixteenth century for watches provided with a "dial to the sun."[2] The document mentions: "To Abraham de La Garde, horologer of the King the sum of one hundred forty five escus for parts made and delivered by him, attending to the person and the service of the King, during the months of January, February and March of the year one thousand five hundred ninety-one. To Wit:

For a cut silver watch, gilt with a dial to the sun.
Here . . . . .35 escus
For a small watch in a completely cut and gilt case with a dial to the sun, with gold enameled hours.
Here . . . . .40 escus
For two other watches in silver cases, also gilt [each] with its dial to the sun, the dial with hours also gilt enameled, at the rate of thirty-five escus each.
Here. . . . . .70 escus"

The *wheels*, made of brass, were molded, then cut out. At that time their teeth were cut through with a file, hence their irregularity. Once cut, the wheels were mercury-gilt to protect them from corrosion. The pivots would rotate in holes "sunk into" the metal of the plates. The movement was made of four wheels: the fusee wheel, the intermediate wheel, the contrate wheel and the crown or escape wheel. The fifth wheel, called the "big intermediate," the "center" or the "minute" wheel, was introduced into the movement only at the end of the eighteenth century.

The motive power of the spring is carried by this wheel and pinion train to a mechanism called the escapement. The *escapement* communicates motive power to the foliot—at first straight, later circular—and maintains its oscillations. Its function is to give a discontinuous movement to the wheel train. In order to do this, it has a system that stops the wheelwork, then lets a tooth "escape," then again stops it, and so on.

The only escapement used in sixteenth- and seventeenth-century watches is the "verge," also called the "crown wheel." It is characterized by an escape wheel, dented like a crown, positioned at a right angle to the other wheels. The axle of the foliot, the verge, is parallel to that wheel. It carries two paddle-shaped pieces called *flags*. The edgewise meshing of the flags with the upright teeth of the crown wheel produces alternating back-and-forth movements which cause the foliot to oscillate. The crown escapement was used until the early nineteenth century. The best horologists of the eighteenth century, such as Pierre Le Roy, Ferdinand Berthoud and John Harrison, praised its qualities, particularly its solidity and the simplicity of its construction.

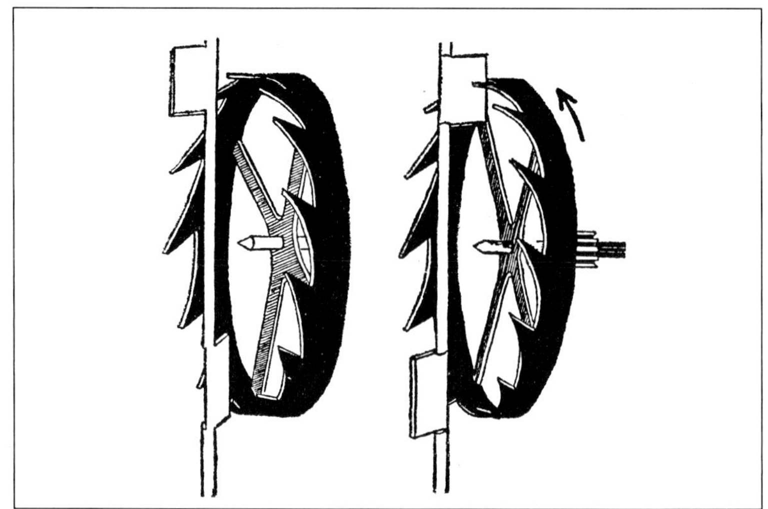

31 Verge escapement.

To confer some degree of exactitude to the watch, it became necessary to add regulating systems to these fundamental parts.

Regulating systems

Regulating a constant driving force

The constant transmission of the motive power of the driving spring to the wheels was the essential condition to the well-running of a watch. Indeed, the crown escapement beats faster when the motive power is strong (wound spring) than when it is weak (unwound spring). To obtain a perceptibly constant driving force, two systems were invented: the stackfreed and the fusee.

The stackfreed mechanism was used by German horologists during the sixteenth century and in the beginning of the seven-

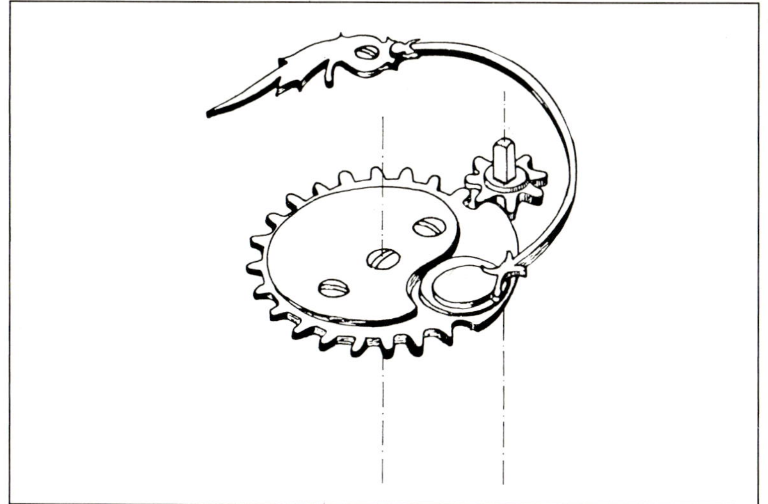

32 Regulating of the motive power of the spring by a stackfreed.

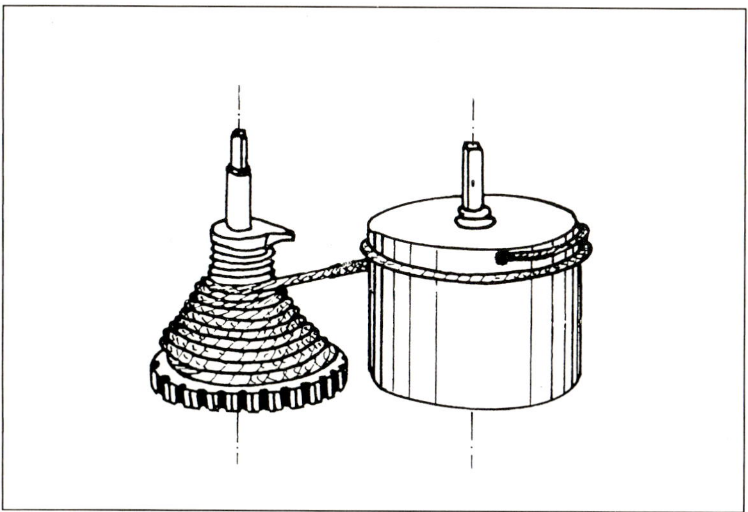
33 Regulating of the motive power of the spring by a fusee.

teenth. A snail-shaped cam is mounted on a wheel meshing with a pinion geared on the barrel arbor. A long, arched, stiff blade spring ending with a friction roller bears against the cam and has a braking effect. The cam makes exactly one complete turn during one run of the movement. When the driving spring has just been wound, the roller presses forcefully against the greater-length radius of the cam; it thus considerably brakes on the movement of the spring, which then tends to unwind. Then, as the effort of the spring weakens, the pressure of the roller is exerted on progressively diminishing radii of the cam, which reduces the braking action, until the roller reaches the concave part of the cam, below which the wheel has no teeth. In this position, the stackfreed lets the driving spring freely develop its last effort.

The system efficiently equalized the force of the power spring. It offered the advantage of the fusee's occupying less height and permitted the building of flatter watches. But by regulating the motive power through braking, it had the disadvantage of consuming some of its energy.

In the eighteenth century, Thiout wrote about the fusee in his *Traité d'horlogerie* (vol. 2, p. 236): "This principle founded on the most natural laws of Mechanics is the most perfect to correct the most irregular motive force."

The fusee is shaped like a truncated cone, grooved all around from bottom to top in a continuous spiral. A gut cord, replaced in the middle seventeenth century by a steel chain, winds round up the groove. After winding, the cord (or the chain) is completely coiled up around the fusee. Hooked to the barrel, it receives the driving force of the spring it is to regulate. When the spring is at its strongest, it pulls on the portion of the spring at the top of the fusee, that is, at its smallest radius. As the spring relaxes its tension, the cord wraps around the barrel and the pull is exerted on a progressively longer radius. When the movement stops, the whole cord is wound around the barrel, except for the extremity attached to the bottom of the fusee. The unevenness of the motive power is thus corrected according to the principle that "a half-smaller force, applied to a radius or a lever twice its length, remains equal to the one that was double on a half-shorter lever."[5] The force of the driving spring is thus evenly carried by the fusee chain to the first great wheel of the train, under the base of the fusee, which engages with the first pinion of the movement.

Gut cord was used until around 1660. Its main disadvantage was that it stretched, particularly under humid conditions. It was replaced by a steel link chain, the invention of which has been occasionally credited to Gruet, a Genevan watchmaker. No horologist of this name has yet been found in any archives.

Regulating the balance wheel

The good running of a watch depended also on the regulation of the balance. Before the revolutionary introduction in 1675 of the regulating spiral spring, also called hairspring or balance spring, various systems were imagined to resolve the problem and transform the old foliot into a true balance mechanism.

The primitive foliot was a simple brass or steel bar bearing at each extremity a small compensating weight, the head of the axle

34 Oval watch. Movement signed "Aug Büschman." Augsburg, about 1600. L. 84; W. 51.
The silver false plate of the dial is engraved, in the style of the period, with flowery foliage and peopled with two squirrels, the face of an angel, and a man sitting on a barrel. An annular dial indicates the time in Roman numerals; a second one permits the regulation of the alarm mechanism.
La Chaux-de-Fonds, Musée International d'Horlogerie, inv. 1147

35 Astronomical round watch. Silver and brass engraved case. Movement signed "G Nourrisson A Lyon." Mid-seventeenth century. D. 50; Th. 23.
Besides the time, the dial indicates the days of the week, the months, the date and the phases of the moon. The movement, possessing an alarm, is the work of a reputed Lyons watchmaker, whose two sons, Guillaume and Pierre, were also watchmakers.
Paris, Musée National des Techniques (CNAM), inv. 20289

upon which it was fixed rotated inside a hole dug into the metal of a bridge called the cock. In order to regulate the length of its oscillations, horologists used *hog bristles*. An early process had been to fix to the plate the extremity of a bristle; in this case, a gudgeon, fixed at a right angle to the foliot, came to bank against the free extremity of the bristle. A second method was to fix the long bristle at the extremity of the foliot; the oscillations of the foliot were limited by two buffers between which the distance could be modified.

Around the mid-sixteenth century, a circular foliot, whose shape was similar to the one of the future balance wheel, replaced the straight foliot. It was then thought the oscillations would be regulated by a lever carrying *movable pins*, positioned between the arms of the foliot. Displacing the pins toward the arms of the ring, the oscillations were wider, and consequently the watch ran slower. Alternatively, if the pins were closer to the central pivot, the amplitude diminished and the watch ran faster.

Complications

The inexactitude of primitive watches did not prevent watchmakers from adding complications to their movements, such as striking "*au passage*," alarum bells or astronomical indications.

Striking watches

As early as the start of the sixteenth century, striking watches formed a large part of the horologists' production. The inventory in 1534 of the inheritance of Francis I's home secretary, Florimond Robertet, included a dozen watches, seven among which had a striking mechanism.[4] In 1579, Maurice-Bernard Ferry, a watchmaker to King Henry of Navarre, repaired two watches with bells belonging to Queen Marguerite de Valois. The document states: ". . . the sum of ten escus is ordered as payment to him of the parts that follow, to wit: to have mended, cleaned and polished a large watch belonging to the said lady and rebuilt three new wheels for the striker; the spring of the alarum along with three other parts in the movement of one of the said wheels, the sum of six escus. To have mended another small watch of the said lady, ornamented with diamonds and rubies, and have made for it a new main spring, repinioned the striker and cleaned all the movements of the said watch, as much as to putting new cords, the sum of two escus. . . ."[5]

The infatuation for watches striking by themselves, "*au passage*," lasted during the whole seventeenth century until the time when the new repeaters were preferred. In the archives in Blois, Abbé Develle found many of them mentioned: a "ringing watch" offered by the inhabitants to Anne of Austria to mark her visit in 1616 and another one, "striking on each hour" was pointed out in P. Cuper's shop.[6]

Because of their striking, which could happen in untimely moments, these watches could expose their owners to embarrassing situations. In *The Liar*, a comedy written in 1642, Corneille gives an amusing example (Act II, Scene 5). Dorante has managed to hide himself upon the arrival of the father of his beloved. The father is about to leave when our hero's watch starts ringing.

The good man was leaving, when my watch struck its bell
And he, turning around to his aghast daughter: "Do tell
Whence the watch? And who gave it to you, my child?"
"Acaste, my cousin sent it anon," she then declared,
"He wants to have it cleaned, and its movement repaired"
"Hand it to me", he said, "I'll see that it is done."
She goes to my corner, as I hand her my watch
And lo! To its cord, my pistol does catch
Starts the trigger and bang, a shot is gone!

Alarum watches

The first regulation of the Horologists' Guild in Geneva, promulgated in 1601, assigned as a masterpiece "a small alarum clock hanging from the neck." The statutes of the Paris Community, approved in 1646, also required that kind of work, difficult to execute, to the candidates for mastership.

This added feature was already current in the sixteenth century, as attested by, for instance, the already mentioned repair made by Ferry in 1579. Two alarm systems were used by watchmakers in the seventeenth and eighteenth centuries. Most often, the moving alarm dial, located within the circle of the hours dial, presented the Arabic numerals 1 to 11, arranged in the reverse order of the Roman numerals on the hours dial. A steel pointer, located on numeral 12, had the function of an hour hand. The alarm hand would start the strike when it reached XII. To be awakened at the desired time, it was sufficient to set this hand upon the anticipated hour indicated on the alarm dial. The disk turned along with the hand until it reached noon (or midnight), corresponding to the chosen time. For instance, if one wanted to wake up at VIII, one would move the hand above 8 of the small revolving dial. When the pointer indicated VIII o'clock, the hand was on XII and the hour would be struck.

There was another system, slightly different and less often used.

The Arabic numerals of the revolving dial were engraved in the same direction as those on the hours dial. There was no pointer on 12. A big hand with its counterweight marked the hour. The numeral on the alarm dial, corresponding to the hour one wanted to be awakened, was placed under that hand. When the hour hand indicated this hour on the dial, the bell was struck. At that moment, the numeral 12 of the alarm dial was always in front of VI on the hours dial.

Watches with astronomical indications

In the days of the Renaissance, astronomical questions were more than ever topical issues. The reproduction, through mechanical means, of sidereal movements was a problem interesting both scientists and princes. Astronomical clocks were then being ordered by Charles V, Charles of Lorraine, Albert of Brandenburg, Augustus of Saxony and William IV of Hesse. Generally, they indicated—besides the mean time, the apparent solar and sidereal time—the movements of the sun and the moon, the date, the days of the week.[7]

Watches showing astronomical indications, to which at times was added astrological information, were also created from the start of the sixteenth century. The inventory in 1534 of Florimond Robertet's possessions mentions a big gilt brass watch he had ordered "that marks all the stars, all the celestial signs and movements, which he understood very well."

All through the seventeenth century, astronomical watches were greatly in favor. They seem to have been a specialty of Genevan horologists, if we are to trust the preserved pieces. Their movements, indeed, bear the signatures of Jacques Sermand, Jean-Baptiste Duboule, Henry Ester, Antoine Arlaud and Pierre Duhamel. Their hour dial is off-center in order to leave room to other dials indicating the months, dates and zodiacal signs and to apertures where the days of the week, with their symbols, appear, as well as the age and the phases of the moon.

Despite the ingenuity of the horologists of the sixteenth century and the first two thirds of the seventeenth, the watch was still an imprecise horary instrument, unable notably to solve the problem of longitudes at sea. Fortunately, horological technique profited by the scientific upsurge of the seventeenth century. A series of researches undertaken by the ablest scientists of the time reached its apex in 1675 with the achievement by Christiaan Huygens of the balance-regulating spring spiral. From then on, the conquest of precision was on its way.

36 Plate models drawn by Marin Estienne in his manuscript *Des machines de montres et horloges,* 1693 (fol. 30).
We note that M. Estienne presented numerous models in a very traditional style, like these, equipped with a one-footed cock and a wormscrew.
Caen, Musée des Beaux-Arts, Mancel Coll., manuscript 253

II "Small-work" horological research and its applications from 1650 to 1765

The precision of watches and pendulum clocks depends on the application of a combination of rules, the improvement of which is largely due to a better knowledge of mechanics. In addition, the development of that science during the seventeenth century brought exceptional progress in the domain of horological technique. L. Defossez remarked on the subject: "For mechanics and for horology, the seventeenth century was also the Great century."[8]

The second half of the century saw the appearance of the two fundamental inventions that brought to timepieces a precision until that time impossible to achieve: the pendulum in clocks and the spiral spring in the balance of watches, both discoveries of the mathematician Christiaan Huygens. Various research, undertaken in the same period as the one pertaining to the profiles of dented wheels or the problem of friction, led to significant improvements in watch construction.

By the start of the eighteenth century, the watch movement, to be very precise, required no more than a better escapement and some manufacturing improvements. Research on measuring time, which had deeply interested the scientists in the preceding century, was left to the master watchmakers. The mathematicians of the new century did not provide the theoretical support that would have permitted the successful application of many ideas suggested by the watchmakers. Certainly regretting the situation, the writer of the article "Horology" in the *Encyclopédie* stated: "We owe thanks to these skillful artists for the great amount of research they have done & above all for the perfection of their workmanship; for they [the mathematicians] have not dealt with the theory & principles of the art of time measuring, at all."[9]

Research and inventions

Scientists' participation during the seventeenth century

To improve the exactness of watches, scientists invented new mechanisms. Thus, Leibniz conceived of an entirely new watch works arrangement, which he described in the March 25, 1675, *Journal des Sçavans* (pp. 93–96). But he was disappointed in not finding a master watchmaker audacious enough to realize his idea. Leibniz had replaced the traditional balance by a dented wheel leading a flier whose movement was obtained by the slackening of two springs alternately cocked by the drive spring. The Englishman Robert Hooke (1635–1703), one of the most eminent scientists of his time, created a watch with two balance wheels meshing with each other and moved by the same escape wheel, then, a watch whose balance wheel was activated by a magnet. He also invented, around 1670, the first wheel-cutting machine.

Perfecting the shape of toothed wheels was a problem that sparked much research. The Danish astronomer and physicist Olav Roemer (1644–1710) arrived in Paris in 1672 and continued the work of Girard Desargues (d. 1661), who had been the first with the idea of epicycloidal toothed wheels. The academician Philippe de La Hire (1640–1718) also took part in that research. Volume IX in the *Histoire de l'Académie royale des Sciences* (1666–1699), entirely dedicated to his work, contains his *Traité des épicycloïdes et de leurs usage dans les machines*. This researcher's desire to improve horological science was manifested in several memoirs published by the Académie des Sciences. Volume IX, already mentioned, also offers a study on the "shape that must be given to the fusee of clocks that use a spring as the principle of the movement."

The interest of seventeenth-century scientists mainly solidified around the problem of balance regulating. Before C. Huygens's outstanding invention, several attempts had already been made to solve the problem. Blaise Pascal and the duke of Roannez tried in 1660, with the help of watchmaker Martinot, to apply a spring to the balance. Meanwhile, the same matter was holding Robert Hooke's attention. He was hoping to regulate the balance with a straight spring. On July 7 1674, Abbé Jean de Hautefeuille presented a report to the Académie Royale des Sciences on the possi-

37 *Christiaan Huygens* (1629–1695), portrait by G. Edelinck. Paris, Bibliothèque Nationale, Cabinet des Estampes

Christianus Hugenius, Zulichemius, Constantini Filius,
Mathematicus Celeberrimus,
Denatus Hagæ-Comitum, in Patria, Die VIII Julii An. 1695. Ætat. Suæ, LXVI.

64 JOURNAL

tan & du miniſtere du Grand Viſir. Il en promet la ſuitte avec celle de ſon voyage de Grece; & cependant le Sieur Guillet ſon frere travaille à un plan de l'ancienne Grece enrichi de quantité d'autres découvertes qu'il a faites dans l'hiſtoire Grecque.

DISSERTATIONES ACADEMICÆ DE Oratoria Hiſtoria & Poëtica. In 12. A Paris, chez Michel le Petit.

LE ſecret de plaire ſans art, eſt le plus bel art du monde, & ce n'eſt pas toûjours le moyen le plus ſeur de plaire que de s'attacher ſcrupuleuſement aux regles de l'art. Le P. Olivier ayant pris garde à ce qui plaiſoit le plus à ſes Auditeurs en prononçant la pluſpart de ces Diſcours Academiques, l'a obſervé dans tous les autres. Les ſujets en ſont curieux; comme celuy des larmes, de ce qu'il y a de plus admirable dans la Poëſie, & du caractere du Panegyrique qui eſt l'écueïl des Orateurs. La Latinité en eſt pure, le ſtile net, & les vers dont il les accompagne quelquefois, ont cela de commun avec ceux des meilleurs Poëtes, qu'ils ſont d'autant plus travaillez qu'ils ſemblent ne l'eſtre pas.

EXTRAIT D'VNE LETTRE DE Mr. HVGENS à l'Auteur du Iournal, touchant une nouvelle invention d'horloges tres-juſtes & portatives.

AYant trouvé une invention long-temps ſouhaittée, par laquelle les horloges ſont renduës

DES SCAVANS. 69

renduës tres-juſtes enſemble & portatives; je crois que ce ſera faire choſe agreable au public de luy en faire part. C'eſt pourquoy je vous envoye la deſcription & la figure du modele, qui contient ce qu'il y a de particulier dans cette invention; afin que parmy d'autres nouveautez en matiere de ſciences, vous puiſſiez, s'il vous plaiſt, les inſerer dans voſtre Journal.

Les Horloges de cette façon eſtant conſtruites en petit ſeront des montres de poche tres-juſtes, & en plus grande forme pourront ſervir utilement par tout ailleurs, particulierement à trouver les longitudes tant ſur mer que ſur terre, puiſque leur mouvement eſt reglé par un principe d'égalité, de meſme qu'eſt celuy des pendules corrigées par la Cycloïde, & que nulle ſorte de voiture ne peut faire arreſter.

Le ſecret de l'invention conſiſte en un reſſort tourné en ſpirale, attaché par ſon extremité interieure à l'arbre d'un balancier equilibre, mais plus grand & plus peſant qu'à l'ordinaire, qui tourne ſur ſes pivots; & par ſon autre extremité à une piece qui tient à la platine de l'horloge. Lequel reſſort, lors qu'on met une fois le balancier en branle, ſerre & deſſerre alternativement ſes ſpires, & conſerve avec le peu d'aide qui luy vient par les roües de l'horloge, le mouvement du balancier, en ſorte que quoy qu'il faſſe plus ou moins de tour, les temps de ſes reciprocations ſont toûjours égaux les uns aux autres.

1675. S

bility of regulating the balance with a spring, without specifying the shape it was to have. The idea was "in the air," but none of the researchers indicated the spiral shape of the spring that Huygens was about to imagine.

A revolutionary invention: the regulating spiral spring (1675)

Before the discovery of Christiaan Huygens (1629–1695), the absence of a true regulator blocked all progress toward precision in the watch. The foliot did nothing at that time to improve the oscillatory movement, to which it opposed its inertia. Hog bristles eased the return of the foliot but did not give it any true periodicity.

Alternately, the spiral-shaped spring, attached to the plate at one end and at the other to the balance wheel spindle, gives the latter an oscillatory movement with a proper, even period. Like a clock's pendulum, the regulating spiral makes concrete the idea of the isochronism of oscillations that Galileo had enunciated.

The application of the spiral spring to the balance wheel immediately improved the exactness of the watch. To regulate the running of the movement, it is enough from then on to modify the spiral: its length is increased if the watch runs too fast because of overly rapid oscillations; it is decreased if the watch is slow because of overly slow oscillations.

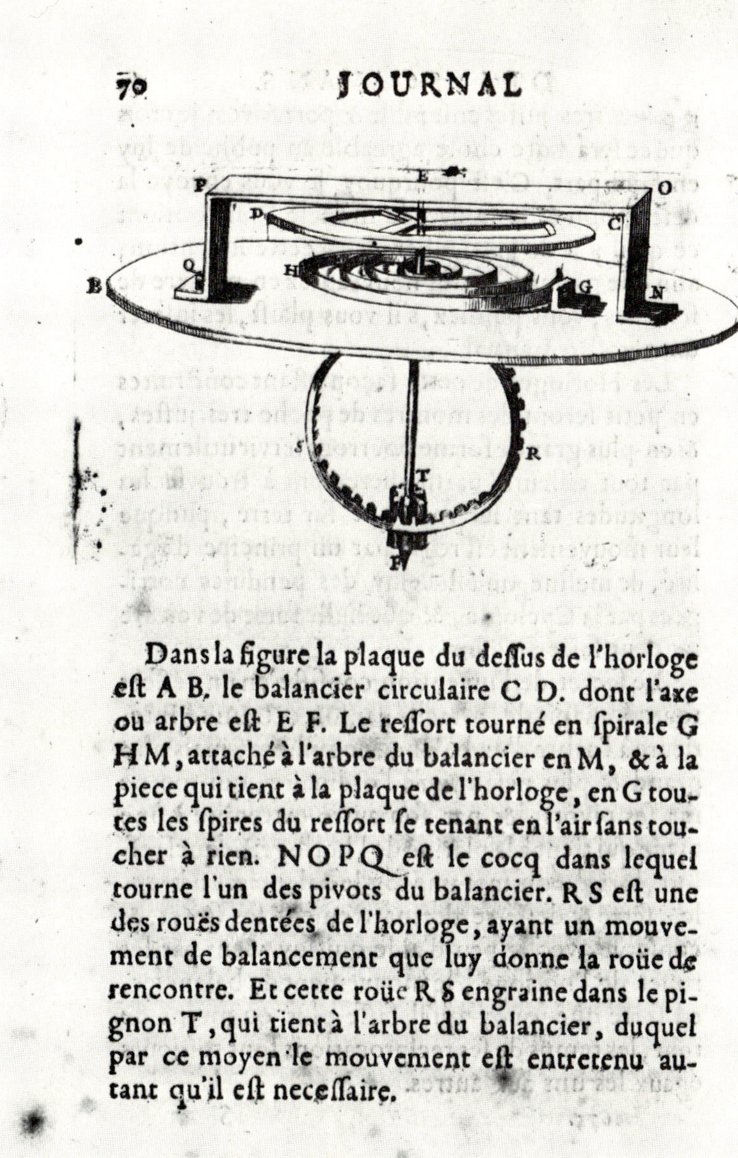
70 JOURNAL

Dans la figure la plaque du deſſus de l'horloge eſt A B. le balancier circulaire C D. dont l'axe ou arbre eſt E F. Le reſſort tourné en ſpirale G H M, attaché à l'arbre du balancier en M, & à la piece qui tient à la plaque de l'horloge, en G toutes les ſpires du reſſort ſe tenant en l'air ſans toucher à rien. N O P Q eſt le cocq dans lequel tourne l'un des pivots du balancier. R S eſt une des roües dentées de l'horloge, ayant un mouvement de balancement que luy donne la roüe de rencontre. Et cette roüe R S engraine dans le pignon T, qui tient à l'arbre du balancier, duquel par ce moyen le mouvement eſt entretenu autant qu'il eſt neceſſaire.

38 a–c. Letter by Christiaan Huygens and addressed on 1675 to the *Journal des Sçavans,* about the regulating spiral spring, accompanied with a drawing.

Huygens announced his invention to the public in a letter published in the *Journal des Sçavans* (1675, pp. 68–70) and then in Volume X of the *Histoire de l'Académie royale des Sciences* (1666–1699, p. 549): "Having found an invention long wished for, one by which the Clocks are made very exact & portable, I believe the public will be glad to be apprised of it. . . . Small Clocks of this sort will be very precise pocket Watches, & larger ones will usefully serve in other ways, particularly to find longitudes, as much on the sea as on the land, since their movement is regulated through a principle of equality. . . . The secret of the invention lies in one spring's being coiled in a spiral, attached at its inside extremity to the arbor of a balance wheel, but larger and heavier than usual and turning on its pivot, & at its other extremity to a part holding the plate of the clock. Once the balance is set in motion, this spring alternately tightens and loosens & keeps the movement of the balance, slightly assisted by the wheels of the Clock, in such a way that whether it makes more or fewer revolutions, all its alternating movements are equal."

Before revealing his invention to the public, Huygens ordered, on January 22, 1675, from Isaac Thuret, a King's watchmaker, a model conceived on his instructions. On January 30, 1675, the scientist announced his revolutionary project to the secretary of the Royal Society in London by an anagram that, unscrambled, meant: "*axis circuli mobilis affixus in centro volutae ferrae*" [The arbor of the mobile ring is fixed at the center of an iron spiral]. The next day he presented to Colbert the model made by Thuret and requested a French patent for his invention on February 5.[10]

From the first steps he undertook toward gaining official recognition of his discovery, Huygens met with difficulties. Indeed, Thuret had made a second model, which he had already presented to Colbert on January 23 as his own invention. Finally, he had to confess his attempted expropriation and Huygens got the patent he had claimed. But Abbé de Hautefeuille vehemently opposed the registration of the patent letters of Louis XIV, claiming that Huygens was himself a thief. His opposition was in vain, and the royal privilege was registered. Nevertheless, the fight helped discourage the scientist, who renounced his patent and returned to his native Holland. He wrote in his *Diary*: "Leaving Paris on July 1, 1676, for Holland; after a five-month illness, I give all watchmakers absolute freedom to work on the invention, seeing what the privilege has already cost me in appeals to have it registered in Parlement, and that after all this I would certainly have lawsuits and new annoyances."[11]

In England the announcement of his invention had been surrounded with the same climate of controversy. Hooke had the grant of the English patent suspended. However, the discovery made a lot of noise, and its application rapidly became widespread.

A small, unpublished treatise dated 1693, *Des machines de montres et horloges. De leur invention et de leur progrès et perfection où on les voit aujourd'huy* (About the machinery of watches and clocks. About their invention and progress and the perfection in which they are seen today), gives us an interesting insight on the new watches.[12] The author, Marin Estienne, a watchmaker established in Caen, devoted the end of his study to their representation, in several remarkably precise drawings on which the names of the

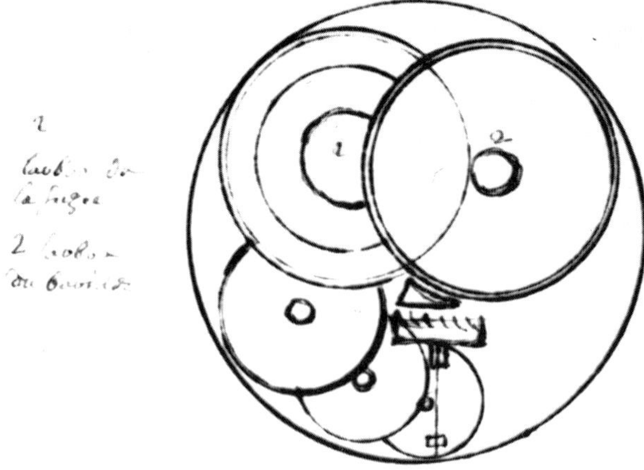

39

40

39 Model of a "spiral Watch, the arrangement of which is among the most pleasant," by Marin Estienne, *Des machines de montres et horloges,* 1693, fol. 39.
With a few drawings, the watchmaker presented in his treatise the novelties of his time. Notice the signature, "Martinot A Paris." Caen, Musée des Beaux-Arts, Mancel coll., manuscript 253

three most eminent watchmakers of the time, Gribelin, Thuret and Martinot, can be read. The title of that part of the book tells us the name given at that time to the new watches: "Arrangement of Spiral Watches, Improperly Named Pendulum Clocks."

Though the exactness of the watches had been considerably improved by the regulating spiral, the conquest of precision still required the solution of problems caused by the friction of pivots and changes in temperature.

New escapements

The problem of the escapements is certainly the one that caused the largest amount of work and research by the eighteenth-century master watchmakers. In this field the pioneers were the English horologists. Thomas Tompion (1639–1713), George Graham (1673–1751) and Thomas Mudge (1715–1794) established the essential principles on which the research on escapements developed. French watchmakers started to take part in the undertaking during the first third of the century. Many escapement models were presented to the Académie Royale des Sciences.

The horologists' research mostly bore on dead-beat escapements; long and difficult, this research often brought no satisfactory results. Invented to maintain the balance oscillations with the most possible precision and thus to ensure the exactness of the watch, the dead-beat escapements were characterized by a part interdependent with the balance, which moves without resistance

40 Round watch. Movement signed "Martin Gerdts Hamburg." Around 1680. D. 50.
The movement offers an interesting example of the efforts made by watchmakers to improve the precision of watches, thanks to Christiaan Huygens' inventions. M. Gerdts used a short pendulum whose oscillations' amplitude can be modified by regulating the two spring buffers (one of them is missing).
Kassel, Hessisches Landesmuseum, inv. U 7

on the arcs described by the escape wheel until it meets the tooth, which gives it a new impulse.[13]

The first dead-beat escapement was achieved in 1695 by Thomas Tompion. It was composed of an escape wheel formed by circular arcs and a notched roller. The impetus took place during only one out of two balance oscillations. It is what is called the "*à coup perdu*" (coupe perdue) escapement. Its main failing came from the excessive duration of the resting pause of the tooth against the cylinder. It was used, more or less successfully, until the end of the nineteenth century, particularly in ordinary watchmaking.

Pierre and Jacques Debaufre, French watchmaker refugees in London, presented a dead-beat escapement in 1704 whose pivots revolved on pierced rubies. Newton was able to confirm the precision of the watches provided with that escapement. Both watchmakers had tried to combine the traditional system of the verge escape wheel with the new principle of the dead-beat escapement. The escape wheel was formed by two superimposed ratchet-tooth wheels in such a way that the teeth of one were staggered in relation to those on the other wheel. The plane of the new escape wheel was parallel to the axis of the balance, whereas in an escapement with the traditional crown wheel, it is the profile of the wheel that is parallel to the axis of the balance wheel. Each wheel alternately presents a tooth, which pauses on the curved surface of the cylinder, then impels the mechanism by sliding on the inclined plane presented by the cylinder.

This type of escapement was used with some modifications by the English watchmaker Henry Sully, who had been attracted in France by the regent. Pierre Le Roy, the brother of Julien Le Roy, was also interested by the De Baufre escapement and presented a few corrections to the Académie des Sciences in 1742: "Mr. Bauffre, a French watchmaker established in London, imagined in 1704 the Escapement that is named *à repos* [dead-beat], in which he suppressed the palets & the crown wheel & put in their stead a kind of cylinder & two other wheels. . . . Mr. Le Roy substitutes for Mr. Bauffre's cylinder a small truncated cone & for his two wheels, which are flat, two horizontal wheels that impel the balance wheel. The first watch he executed on this scheme was finished in 1737 & served as example & proof for a considerable wager that had been made in Lisbon over preferability of the watches of England and France. This one compared so well with a watch of the famous Mr. Graham, that it was impossible to decide which was the best. . . ."[14]

The cylinder escapement is a successful variant of the dead-beat escapement. It was thought out around 1725 by George Graham, a student of Tompion. Graham's escapement was formed with a hollow steel cylinder and a brass escape wheel. It was in use for

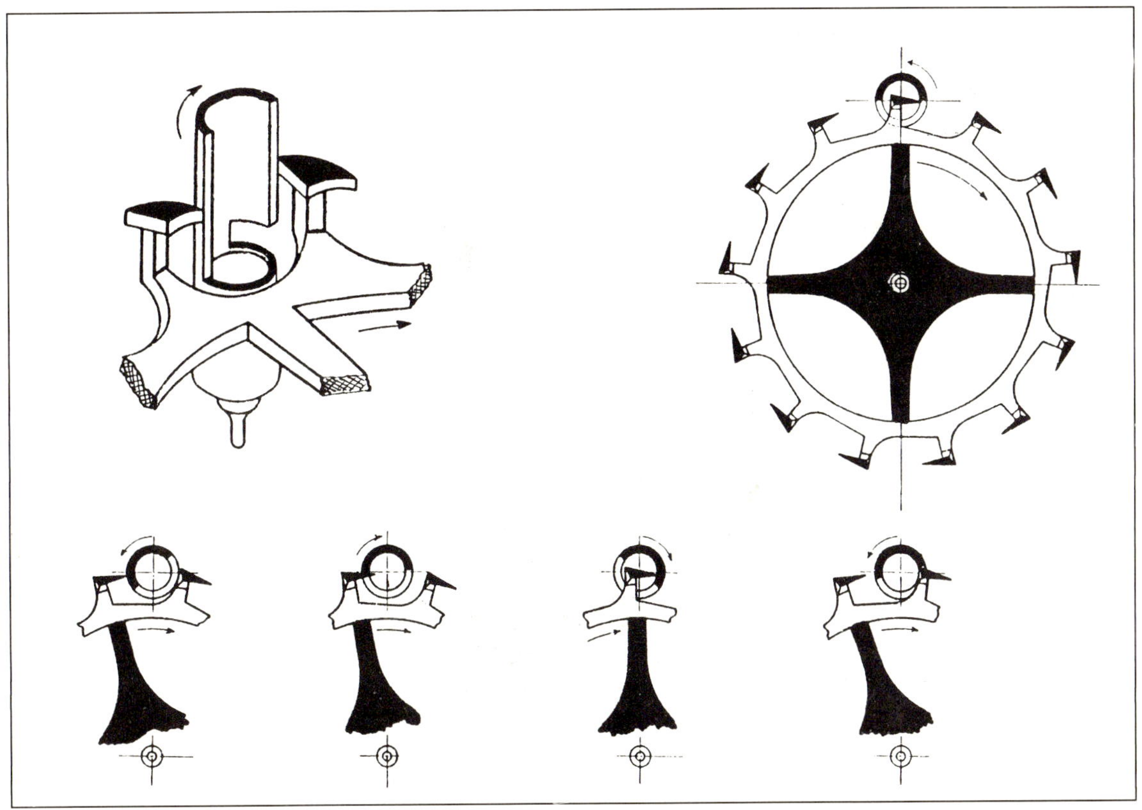

41 Cylinder escapement

around fifty years with few notable modifications. Before the years 1760–1770, watchmakers seldom used it because of the difficulty of its execution. Later, its adoption was partly determined by the vogue of flat watches, the manufacture of which it facilitated.

The *Mercure de France* gives an account of a significant dispute in the midcentury, opposing Pierre Le Roy, brother of Julien Le Roy, and Godefroy, watchmaker to the late duke of Orleans. The affair started with a letter of Pierre Le Roy to a certain Mr. N. of the Bordeaux Académie des Sciences on the construction of a watch that had been presented on August 18, 1751, to the Royal Academy. In the letter published in the *Journal de Trévoux* (June 1752, pp. 1300–1314), Pierre Le Roy recommended a way to correct the defects of the flat watches imposed on watchmakers by fashion. Before anything, he was against the use of the cylinder escapement, the failings of which he underlined: "Watchmakers will always prefer to make the escapements of their watches with the scape wheel rather than making them with Mr. Graham's dead-beat escapement . . . because, everything being considered, the dead-beat escapement, however ingenious, does not give them, during the running time, more exactness, & even gives them less, than the scape wheel; the friction of Mr. Graham's being so strong, in relation to the weak power of the spiral spring . . . one is forced to oil this escapement abundantly to diminish friction and variations. As long as the oil keeps its even quality, they run more exactly than watches with the escape wheel, but as soon as the oil gets stickier or dries up, that exactness is destroyed; because, the resistance of friction of the cylinder against the ratchet tooth being increased, the resistance varies in relation to the increase, which sometimes reaches the point that the spiral, whose function is to restore to the balance the equivalent of the force it received from it, can no longer overcome the resistance of friction; & that restoring being no longer possible, the movement of the balance ceases, by consequence of which the watch stops. I will add to those failings that the escapement in question is much longer & more difficult to make than the one of the scape wheel; consequently, the Watches that have those escapements have less of a market because of their high price, which is unavoidable."

P. Le Roy recommends the use of the escape wheel in his letter and formulates two conditions:

1. The escape wheel must be bigger than usual to give more solidity to the escapement.

2. The pinions must have seven teeth instead of six: the pinion wings of those with seven acting on the teeth of the wheels with a smaller angle, the friction on the teeth and that of the pivots on the sides of their holes are thus decreased.

Godefroy answered P. Le Roy in the October 1752 issue of the *Mercure de France*. He belittled the construction proposed by Le Roy and, in return, praised Graham's cylinder escapement, denying the inconvenience of the friction noted by Le Roy. According to him, the cylinder watches were easier to regulate, were not disturbed as much by shocks or falls, and could be made as flat as desired without having to change their construction. Letters about the quarrel were published in the *Mercure de France* until July 1754.

In 1752 Beaumarchais, then known as Caron *fils*, presented a dead-beat escapement he called "a double virgule." His discovery was contested by several watchmakers—Lepaute, Biesta and Romilly. Finally, the Académie declared the future author of the *Barbier de Seville* the winner. The escapement is described in the list of the *Machines ou inventions approuvées par l'Académie en 1754*: "A new dead-beat Escapement, presented by M. Caron *fils*. The escape wheel, which is flat & provided with pegs, alternately placed on either side of its plane and perpendicularly to that plane, passes between two crescent-shaped pallets that are united to each other and to the rest of the balance verge by a peduncle similar to the elbow of the crank of a pump. Two parts can be distinguished in each pallet, one dug as a cylindric gutter following the axis of the balance, the other one straight or curved according to the taste of the Watchmaker. The pegs, placed on either sides of the wheel plane, rest alternately in the cylindric gutters of both pallets, & then escape from these gutters, pushing the rest of the pallets, which maintains the vibrations of the balance. Finally, the escape wheel is cut like the others, between its pegs in such a way that the peduncle or crank elbow may lodge itself inside the cuts, & the excursions of the balance may be longer. This escapement has been considered the most perfect to be, until now, adapted to watches, but also the most difficult to execute."[15]

In 1724 J. B. Dutertre showed the Académie a watch with two balances geared to each other. The idea of this double escapement was taken up again in 1754 by Jodin.

In 1748 Pierre Le Roy, the son of Julien Le Roy, presented his detent escapement. The commissioners' report of September 6, 1748, describes the new escapement: "We examined, on the command of the Académie a new dead-beat escapement, presented by M. Le Roy *fils*. A flat wheel on whose plane is ordered a crown of peg-shaped teeth, cut as triangular prisms, catches, as it passes, a single curved pallet attached to the balance arbor and makes it run through a certain arc: when the tooth of the wheel leaves the pallet, another tooth is instantly engaged in a small fork placed at the end of an elbow lever, affixed on the watch plate; & in this lies the main advantage of M. Le Roy's escapement: for, by this means, the balance brought back by the spiral spring is perfectly free in its return & is not impeded by the friction caused by the pressure of

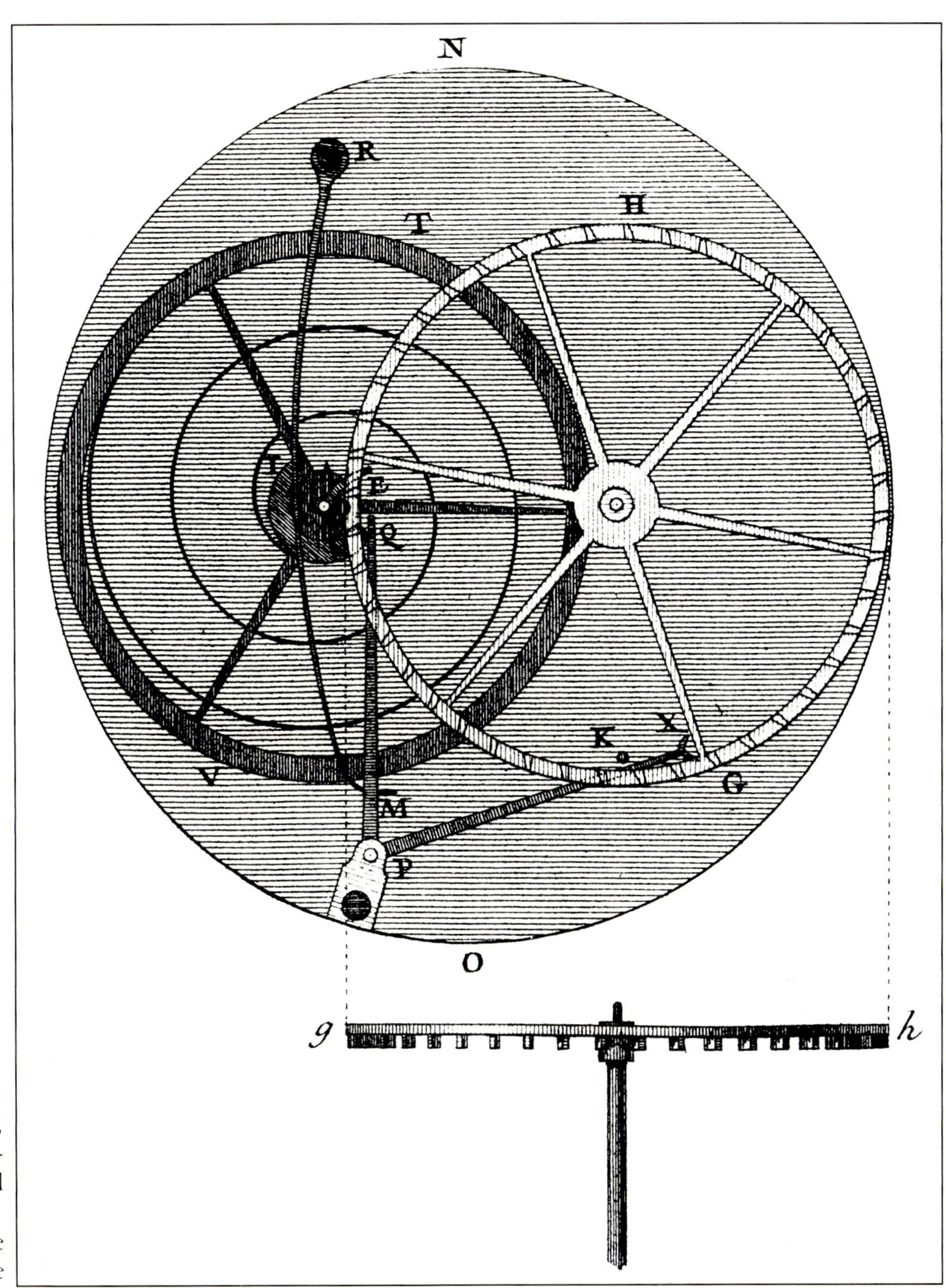

42 Detent escapement invented by Pierre Le Roy, approved in 1748 by the Académie Royale des Sciences. The engraving illustrated the report presented by its author.
NO is the cock plate; *TV*, the balance; *HG*, the escape wheel; *RM*, the spring; *QPX*, the elbow lever; *AE*, the curved palet.

the escape wheel tooth, which is unavoidable in the rest period of ordinary escapements, – this also has the effect of lessening the wear and tear on the tips of the teeth. In this particular escapement the wheel rests for a half-vibration on a part that does not move & is unrelated to the balance. At the end of the return of the balance, the pallet encounters another tooth & makes it recoil a little, which permits the lever pushed by a spring to disengage & the wheel to catch the pallet anew, to make the balance start a second vibration. The idea appeared good to us, and deserves to be followed by its Author, who is in the condition to draw the best advantage from it one can hope."[16]

The repeating mechanism

The "on demand" striking train, commonly known as the repeater was simultaneously invented by two English watchmakers, Edward Barlow and Daniel Quare. Each presented a watch with this mechanism to James I, who finally gave his preference to Quare's. Ferdinand Berthoud relates in his *Histoire de la mesure du temps* the differences between the two watches: "Repetition occurred in Mr. Barlow's watch, by pushing in two little parts, one on either side of the watch case; one repeated the hours and the other the quarters; but Mr. Quare's repeated by the mean of a single pin, located next to the pendant of the case, which, being pushed inside, made the repetition of hours and quarters, as it is done presently by pushing the *pendant* which carries the pin just once."[17]

The repeater is essentially distinguished from the "*au passage*" striker by the fact that it can repeat as one wishes it to, whether hourly or quarterly. To effect this all that is required is to press a button on the head of pendant. The mechanism is mainly made of "snails" and racks. The hour snail makes one turn in twelve hours; it is stepped in twelve plateaux or degrees, each corresponding to the number of times the bell is struck to indicate each hour. When the repetition is started, the arm of a rack dips in on one of the plateaux of the snail and determines the number of times the hammer must strike the bell. A second snail indicates the number of minutes that have to be struck.

Repeaters had a tremendous success throughout the eighteenth century. They were particularly appreciated for giving the time in darkness. Watchmakers strove to perfect the system by adding a chime on the half-quarter hour, every five minutes, or even on the minutes themselves. A part preventing the chime if the button was not correctly pressed in, gave the name *tout ou rien* (all or nothing) to the repeaters that had it.

Julien Le Roy developed repeaters without bells. In this *à toc* (dumb repeater) system, hammers strike small masses of metal soldered to the case. Omitting the bell permitted the building of larger and thus more solid movements. Le Roy gave a detailed description of the new construction in the March 1741 issue of the *Mercure de France*. He firmly opposed the "*à l'angloise*" repeaters, that is, paired cases with striking bells that, according to him, ". . . were good when they were big but became defective when they were made small, because a small movement is more difficult to work on, according to the rules of the art, less good, less durable, more difficult to put back in good condition than if they were of a more convenient size. . . . To obviate these faults, inseparable from small repeater movements, I constantly looked for means to make them as large as possible in middle-sized cases; it is with these views and for the same reasons that I strove to entreat the public to overcome its reluctance to bell-less Repeaters."[18] His perseverance was rewarded, since his invention was rapidly adopted by all French watchmakers.

In February 1744 Le Roy presented in the *Mercure de France* another invention in a memoir "on the old form of repeater watch-cases and on a new way to build them." He recommended using the space lost by the bezel mount to lodge the repetition work.

The repeater is often accompanied with an "*à tact*" (touch) mechanism that permits one to learn the hour discreetly. Instead of striking a bell, the hammer strikes a small lever; muffled blows, corresponding to the number of hours and quarter hours indicating the time, are transmitted to the finger by a sliding rod on the framework of the case.

Reduction of friction

The invention of the hairspring had considerably improved the exactness of watches, but there still remained a few problems to solve, such as friction, before one could speak of a true precision.

Christiaan Huygens himself was conscious of the difficulties it caused. In a letter to Oldenburg written July 11, 1675, he remarked: "As to the exactness of these watches (with a regulating spiral), I consider it incomparably greater than that of the ordinary ones, but always less than the one of the pendulum clocks' . . . the difference being that the pivots' friction is more considerable in a small work than in a big one."[19]

The way to pierce rubies and use them as settings for pivots was discovered around 1700 by a mathematician of Swiss origin, Nicolas Fatio, who settled in London to exploit his process in partnership with two watchmakers, Pierre and Thomas Debaufre, themselves of French extraction. The use of pierced rubies improved the wheel work of the watches by reducing friction and consequently the wear on the pivots.

In the short run, and unfortunately for the competition, English watches were the only ones to receive these advantages, for the process for producing stones for watchmaking was kept hidden from the Continental horologists until around 1770. It was only in the first quarter of the nineteenth century that pierced rubies became generally used in France and Switzerland.

Nevertheless, due again to Julien Le Roy's ingenuity, an invention improved the running of Continental watches. Until then, the top balance pivot rotated in a hole dug into the brass of the cock, without going through the metal. The hole quickly became worn out by the steel of the pivot and clogged with the verdigris formed by the oil used to diminish the friction. Julien Le Roy had the idea to pierce through the hole and to place a stainless steel counter-pivot. The use of this steel cockhead piece became general by 1730.

The fight against the effects of temperature

In the seventeenth century many scientists remained skeptical about the effects of temperature on metals. R. Hooke was one of the first to take this phenomenon into consideration: even before the end of the century, he recommended the use of glass hair-springs, as they were insensitive to temperature variations. With the perfecting of thermometers in the eighteenth century, the question of the influence of temperature changes on the running of watches was clearer and provoked numerous studies.

Temperature caused a watch to go slower in warm weather and faster in the cold by 11 seconds per Celsius per day. The steel spiral became soft when the temperature rose and stiffer when it lowered. Watchmakers tried to find compensating systems that would permit the oscillator period to remain constant in a temperature interval between 0°C. and 40°C. In 1755, Julien Le Roy's son, Pierre, invented a compensating balance, a forerunner of the cut bimetallic balance. The rim of the balance wheel was made of two metals, latten-brass and steel, whose uneven dilatation led to a distortion which tended to reduce the variations in the run of the movement provoked by temperature changes.

Experimental watches

The horologists strove throughout the eighteenth century to improve the running of watches and clocks with ingenious mechanisms. In France they presented their inventions to the Académie Royale des Sciences, which, if it found them worthy of interest, would classify them among the "approved Machines," in a list that was published in an annual volume of the *Histoire de l'Académie Royale des Sciences*. Consulting that publication, along with the *Journal des Sçavans*, the *Mercure de France* and the *Journal de Trévoux*, allows us to follow the horologists' efforts to measure time during the period.

In 1716 the English watchmaker Henry Sully offered a watch of a novel construction, by which friction was reduced, thanks to the use of hard stones.[20]

In 1742 Dutertre *fils* presented an equation watch done for the duke of Orléans. The dial had two hands for indicating minutes: one marked minutes in mean time, the other in true time.[21] In the same year, Gourdain proposed a watch built without a fusee.[22]

In 1754 a Saint-Germain-en-Laye watchmaker, Jodin, presented a watch with two balance wheels geared to each other.[23] In the same year, Ferdinand Berthoud was congratulated for an equation watch that marked "the seconds through the center, the hours and minutes of solar time and mean time, the months of the year and the date."[24]

In 1758 Romilly was selling a watch with the escapement invented by Caron *fils* but that he himself had slightly modified.[25]

In 1761 Pierre Le Roy presented an *à tact* watch, made following the instructions of the prince de Conti.[26]

In 1763 J. B. Le Roy communicated a report in which he recommended placing the fusee upside down in the watch.[27] In that year the Académie also approved a new arrangement of parts in a repeater invented by Jean-Antoine Lépine and two watches by Pierre Le Roy. The first one was a watch indicating the seconds, in which the seconds dial was fastened in the cock plate; it could be seen through a window cut into the back of the case. The second was a repeater in which the watchmaker had done away with many parts.[28]

In 1764 the Académie gave its approval to three new watch constructions. One of them, presented by a Parisian watchmaker, Biesta, had an escapement that could be removed without having to dismantle the rest of the watch. A second arrangement, invented by Nioux, had a second hand that neither oscillated nor recoiled. The third watch, presented by Coupson, had neither barrel nor fusee. The motive power of its movement was provided by a straight spring.[29]

The various inventions made at the end of the seventeenth century and during the first two-thirds of the eighteenth in the field of small-work horology were often restricted to high-luxury watches. However, ordinary production, in which the prettiness of the case was usually more important than the perfection of the movement, received important technical improvements.

Changes in movements

The eminent English watchmakers at the end of the eighteenth century contributed to imposing, in their country, a type of watch that, because of its perfection, kept its characteristics throughout the next century. On the other hand, the more indecisive French created at the time a model of watch that was modified in the second quarter of the eighteenth century, as the prestige of French horology reasserted itself.

During the first decades of the eighteenth century, English movements and French movements showed differences that had originated at the end of the preceding century or appeared as French discoveries were made in both kingdoms.

43

44

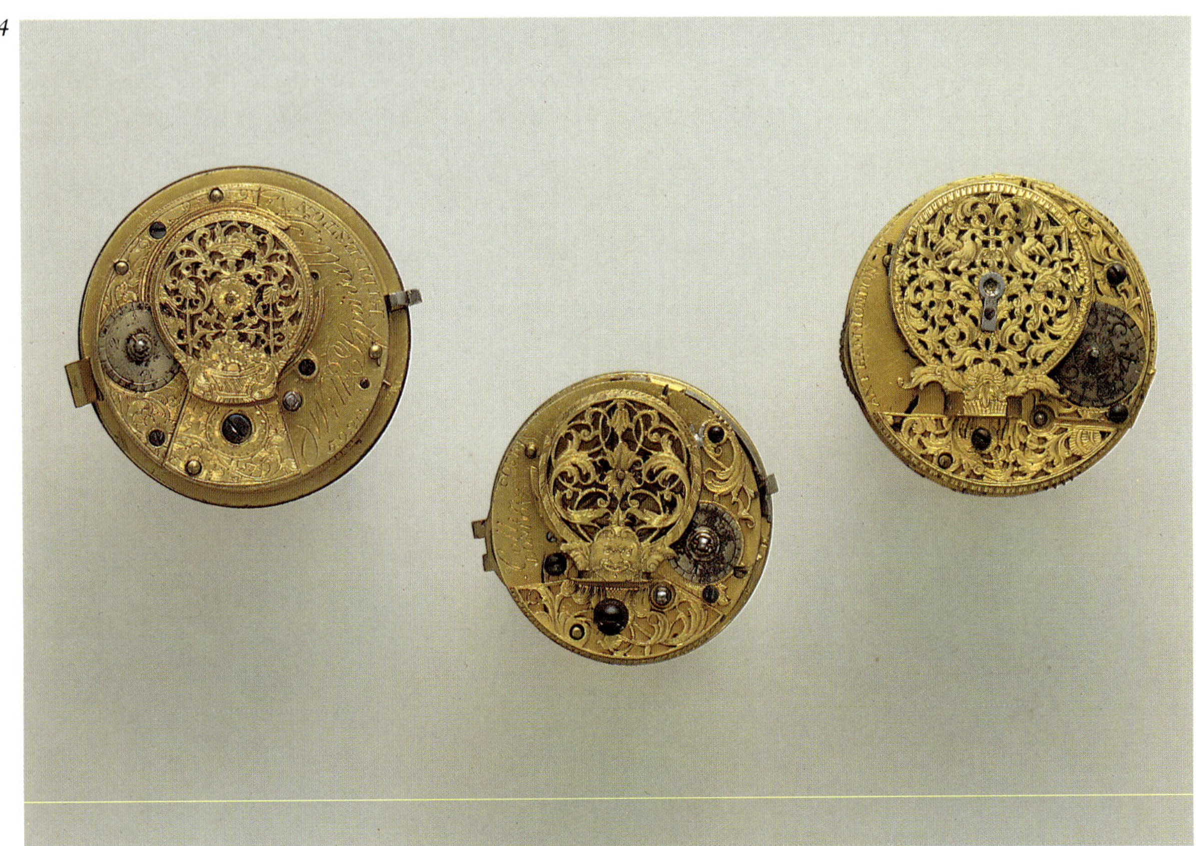

◁ 43 Pair case watch. Engraved gold interior case; sharkskin exterior case. Movement signed "Godfrey C Say London." Mid-eighteenth century. D. 50; Th. 25.
Like most eighteenth-century English watches, this repeater is equipped with a dust-cap.
Paris, Musée National des Techniques (CNAM), inv. 21973

◁ 44 *Left.* Watch movement signed "Willm Squirrell Bildestone 5928." England, end of eighteenth century. D. 44; Th. 13.5.
Pierced cock, engraved at foot.
Middle. Watch movement signed "Collins London." Middle of eighteenth century. D. 38; Th. 13.
One-footed cock, pierced and engraved with mascaron and foliage scrollwork.
Right. Watch movement signed "Antram London." Early eighteenth century. D. 43; Th. 16.
One-footed cock, pierced and engraved with birds and acanthus leaves.
Paris, Musée National des Techniques (CNAM), inv. 10663[1], 10664[4], 10663[7]

45△

46▽

45 *Top.* Round *oignon* watch signed "Gaudron A Paris." Around 1700. D. 42; Th. 20.
The large circular cock, typical of French production between 1690 and 1720, is pierced and engraved with acanthus leaves.
Bottom. Round watch. Silver case. Movement signed "Julien Le Roy à Paris 3773." 1760 (dated letter from the Community House in Paris). D. 45; Th. 25.
The circular cock pierced and engraved with the initials *J L R* characterizes the watches produced by Pierre Le Roy after his father's death in 1759.
Paris, Musée National des Techniques (CNAM), inv. 10664[2] and 19233

46 Watch movement signed "Julien Le Roy A Paris 2781." 1752–1753. D. 40; Th. 15.
The inscription "Inventé par Julien Le Roy en 1740," engraved on the bezel mount, reminds us that the type of construction known as *bâte levée* (raised bezel mount), permitting the lodging of the repeater assembly, is the invention of Julien Le Roy.
Paris, Musée National des Techniques (CNAM), inv. 19247

In general, German and Dutch watchmakers adopted the model developed by the English, while the Swiss were most often influenced by the types of watches created in France.

English watches

Thomas Tompion's work had an important influence on English watch production. At times soliciting the advice of the scientist R. Hooke, he had unceasingly tried to improve the precision of his watches, now possible thanks to the application of a regulating spiral spring. The method he perfected for adjusting and regulating the spiral was used in most eighteenth-century English watches. Several of his researches were resumed and pushed further by George Graham, his brilliant pupil.

English watches essentially differed from the French through their systematic use of the pair case. The same care over protection made use, starting in 1725, of the dust-cap, a metal cup which fit over the movement. This cover above the cock forms a protuberance inside which the watchmaker's name is usually written.

The balance wheel's diameter grew after the addition of the hairspring. The difficulty at the time of making the thin blade that formed the spiral did not permit the blade below a certain force; the mass of the balance had then to be proportionate to that force, and its diameter had to increase, as well as its axis.

This modification brought the enlargement of the cock, formed in English movements by a disk under which the balance wheel moves and a semicircular foot screwed by its heel to the plate. Most high-quality English watches showed, around the mid-eighteenth century, a balance counterpivot made of hard stone, diamond or ruby.

The winding square for English watches is always located on the plate, next to the cock, above the fusee arbor. The dust-cap has a hole at this place, so that without removing it one can wind the watch.

On the plate also appears the regulating system of the hairspring. Tompion's method, that is, increasing or diminishing the length of the spiral spring, was generally adopted. The movement of the pins limiting the outside spire is obtained by the sliding of a semicircular rack geared to a toothed wheel. The wheel, equipped with a winding square, is covered over by a dial whose arbitrarily chosen numbers act as guides to operate the lengthening or shortening of the spiral, done with a key. The wheel is turned until the number that corresponds to the desired regulation reaches a fixed pointer.

A regulating mechanism attributed to Nathaniel Barrow was adopted on a few watches. It is composed of a wormscrew ending in a winding square, maintained by two ring-bolts on the plate. The wormscrew carries a slide on which are mounted two pins that maintain the rectilinear extremity of the regulating spring. The slide is equipped with a cursor that moves along a graduated scale, engraved on the plate.

The improved precision brought by the application of the regulating spiral permitted the introduction of the indication of minutes. Developed by Daniel Quare at the end of the eighteenth century, this innovation was rapidly adopted everywhere. The use of the minute hand required a modification of the arrangement of the movement: the wheel that activated it, going through a turn in one hour, was added in the center.

The best English horologists, such as Thomas Grignion (1713–1784), John Ellicott (1706–1772) and Thomas Mudge (1715–1794), used Graham's cylinder escapement. The use of a smaller balance wheel brought about the diminution of the cock disk.

Watches equipped with a crown wheel were still the most numerous by the mid-eighteenth century; they became smaller under the influence of the fashion of flat cylinder watches.

As in France, repeaters were very much the vogue in England. English watchmakers always built them with a bell.

French watches

In France in the last quarter of the seventeenth century, the use of the regulating spiral brought on the appearance of a new model of watches characterized by their large dimensions and bulging shapes, which clearly differentiated them from contemporary English watches.

The French watchmakers, imitated by many of the Swiss, were convinced that the use of a large-diameter balance wheel with a spiral favored the regularity of the oscillations and consequently the running of the watch. The increase of the balance wheel led to the development of every part in the movement. The plates of this type of watches, known as "*oignons*" on the Continent and "turnips" in England, are often more than four centimeters in diameter, while the pillars are about one-and-a-half centimeters. The considerable volume of the frame permitted a very roomy arrangement of the various parts of the movement.

The balance of the *oignon* is protected by a cock covering over a large part of the top plate. It is circular, but the lateral ears, which receive the two screws fixing it to the plate, give it the appearance of an ellipse. Regulating the hair spring with a numbered "rosette" dial is the most commonly used system.

The winding square of the single-hand *oignon* is at the center of the dial. But the one of the two-handed *oignon* is located on the dial at the level of the number III: the winding is done, in this case, on the square of the fusee spindle. The arrangement has the

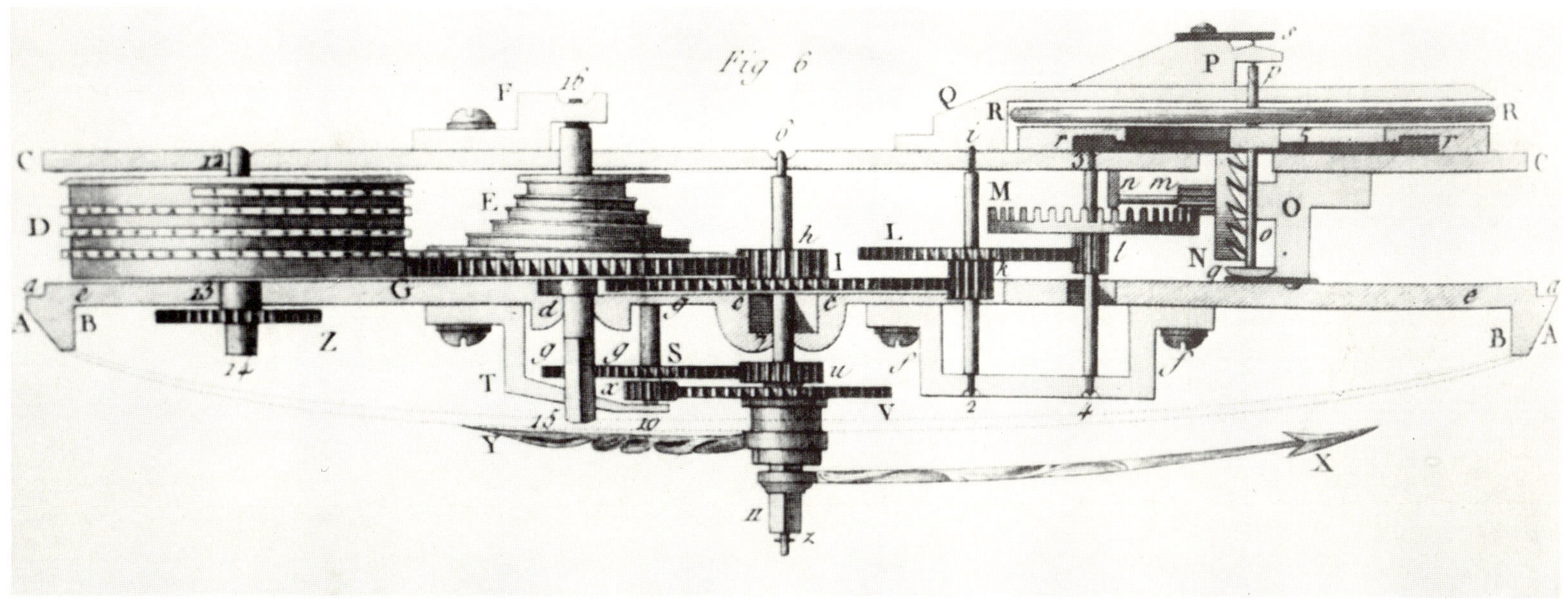

47 Watch movement with two plates and verge escapement. *D:* barrel with chain; *F:* fusee; *R:* balance wheel; *N:* crown wheel; *Q:* cock; *Y:* hour hand; *X:* minute hand.

disadvantage of ruining the enamel of the dial around the opening. French watchmakers introduced the indication of minutes around 1700, a good ten years after their English colleagues.

The dimensions of movements built during the reign of Louis XV tended to diminish. The diameter of the cock seldom reached more than two centimeters. The French watchmakers of that period used the crown-wheel escapement almost exclusively. To respond to the demands of fashion, they built half-flat watches by diminishing the crown wheel. They also manufactured flat watches in which the crown wheel had no more than nine teeth.

The use of the adjustable potence, imagined by Julien Le Roy around 1730, aimed at the improvement of the crown-wheel escape. It rapidly became generally used in quality French watches. The external pivot of the balance wheel entered an adjustable steel part, which could be tightened or untightened with the help of a key or a screwdriver. Thanks to this device, the watchmaker could regulate the action of the escapement more easily.

In less than a century, watchmaking technique had been considerably perfected. The application of the regulating spiral spring, improving, in an immediate and simple way the exactness of ordinary as well as luxury watches, had been the starting point of the research explosion led by the English watchmakers, who were later joined by their French colleagues. At the same time that they brought appreciable improvements in watch construction, they posed the essential problems whose solution would permit a much tighter precision in the running of watches: problems of the escapement, problems of friction, problems of the influence of variations in the temperature. They solved some of them, but it was in the following period that their inventions were perfected and applied to ordinary production. Toward the mid-eighteenth century, the watch still showed many characteristics inherited from its primitive form. Its movement, still lodged between two plates, was invariably equipped with a fusee and generally with the traditional verge escapement. It was the watchmakers of the next period who gave the watch its modern appearance.

III The conquest of precision in the watch, from 1765 to 1820

In the second half of the eighteenth century, the horologists' efforts were concentrated on chronometric research aimed at exactly determining longitudes at sea. Through the medium of marine chronometers, constructed in England and France, combining all the advances that had been brought to small-work horology, the precision of watches greatly improved.

The work on frictional-rest escapements, undertaken by the English and French watchmakers, was continued, while research on anchor escapements, along with the inventions of Jean-Antoine Lépine and Abraham-Louis Breguet, opened a new phase in the history of the watch.

Research and inventions

Marine watches

Scientists and watchmakers had been striving since the beginning of the sixteenth century to discover the instrument that would permit longitude computation. Large rewards had been promised in vain by Henry IV, Philip III of Spain and Louis XIV to the inventor of such an instrument. The invention of the regulating spiral and the many improvements brought to the movements of watches in the beginning of the eighteenth century raised new hopes and new research. The difficulty of such an undertaking can be measured when one realizes that an error of one second brings a 463-meter drift at the level of the equator.[30]

In 1714 the English Parliament encouraged research by instituting a prize of £20,000 to any person who could produce a timekeeping device that could determine the longitude within a half degree, that is, a two-minute approximation, in the course of a crossing from England to India. The prize money was to be reduced to £15,000 if the error came to two-thirds of a degree and to £10,000 if it reached one degree.

Philippe d'Orléans, the regent of France, promised on his part, in a letter addressed on March 15, 1716, to the academician Fontenelle, a prize of 100,000 livres.[31] These were the terms of the promise: "I am sending you, Monsieur, several requests and reports that have been directed to me for some time now by writers from different countries, persuaded that they have found at last the secret, so greatly desired, of exactly and easily determining longitudes. . . . Since before revealing their secret, they insist on assurances of compensation, you can answer them in my name and upon my word that I will have the sum of 100,000 livres be paid to the first person fortunate enough to have found this admirable secret, as soon as the Académie des Sciences has provided me with the evidence. . . ."

After many years of work, in 1761 John Harrison (1693–1776) won the prize instituted by the British government, with his fourth

48 Marine chronometer by John Harrison, 1759 (H4). D. 132.
Because of this watch, in 1761 J. Harrison (1698–1776) won the prize instituted by the English Parliament.
Greenwich, National Maritime Museum, inv. Ch 38/A503 5L

49 Marine clock signed "N° 8 Inventée et faite par Ferdinand Berthoud." Paris, 1767–1768. D. 235; H: 420.
The success in the testing of clocks 6 and 8 permitted F. Berthoud to obtain the title of Watchmaker and Engineer to the King and the Navy. This piece is part of a series of weight clocks—with a gridiron compensation, a circular balance hanging on a blade-spring and rollers—meticulously described by the watchmaker in his *Traité des horloges marines,* published in 1773.
Paris, Musée National des Techniques (CNAM), inv. 1389

marine watch, started in 1758. The watch was equipped with a verge escapement, ingeniously modified, and diamond pallets. The pivots rolled on pierced rubies. A device formed with a steel blade and a latten-brass blade ensured the compensation. When the temperature rose, the dilation of the brass would impose a convex curve on the double blade, shortening the spiral. When the temperature fell, the reverse effect was produced and the double blade lengthened the spiral. After a Jamaica-to-London crossing, Harrison's watch, shipped on the *Merlin*, indicated an error of only one minute fifty-four seconds. It was a triumph. The Board of Longitudes published *The Principles of Mr Harrison's time keeper* in 1767. The same year a French translation appeared. Ferdinand Berthoud was sent by the secretary of the French navy twice, in 1763 and 1764, to examine the watch in London–without success.[32]

The secret of longitudes was a formidable weapon in the hands of a maritime power like Britain. France had to have it also. So in 1765 the French Académie Royale des Sciences proposed, for the year 1767, a prize whose topic was "Determining the best way to measure time at sea." The prize was not awarded; it was proposed

50 Marine chronometer by Pierre Le Roy. Paris, 1766. L. 350.
On August 5, 1766, P. Le Roy presented to Louis XV a clock for measuring the time at sea. This piece is either the one that was presented to the king, named "watch A," or a clock made a little later and named "watch S."
Paris, Musée National des Techniques (CNAM), inv. 1395

51 Marine chronometer signed "John Arnold & Son Inv. et Fecit N° 23." 1790–1791. D. 114.
The movement with fusee and motive spring has a spring-detent escapement (of the type invented by Arnold) and a compensated two-arm (S-shaped) balance with a screw regulating isochronism.
Greenwich, National Maritime Museum, inv. Ch 11/37–1398C

52 Marine chronometer signed "Thos Earnshaw Inv^{t} et Fecit N° 512. London 2856." About 1800. D. 87.
The movement has a twenty-four-hour duration and a spring-detent escapement.
Greenwich, National Maritime Museum, inv. Ch 5/36–605 C

again for 1769. It was won by Le Roy,[33] who had entered two watches. They were tried on the frigate *L'Aurore*, during a sea voyage organized by the marquis de Courtanvaux, assisted by astronomer Pingré. The frigate went only as far as Amsterdam. A new trial of the Le Roy watches was undertaken, under the supervision of Cassini *fils*, during a voyage of the frigate *L'Enjouée*. A new expedition, on the *La Flore*, which lasted from October 1771 to October 1772, permitted verification and comparison of the accuracy of Pierre Le Roy's marine watches and those of his rival, Ferdinand Berthoud. It was recognized that their watches deserved the trust of navigators. The Académie des Sciences offered the same subject again for 1773. There were three competitors: Pierre Le Roy, Arsandaux and Biesta. Ferdinand Berthoud, appointed horologist to the king, had declared that he could not compete since he was the exclusive and salaried purveyor of the navy. Pierre Le Roy again won the prize.

The type of chronometer devised by Le Roy was equipped with a very large steel balance wheel hanging from a metal wire; both pivots of its spindle rolled, each between four cylinders. Two spirals superimposed near the bottom pivot, and proportioned to its size, regulated it. Above them, a metal ring carried two screws that could be tightened or untightened to regulate balance inertia. Le Roy had solved the problem of compensation with two thermometers, consisting of a bulb and an elbow tube filled with alcohol and mercury, placed on either side of the balance. A rise in temperature brought the mercury closer to the spindle of the balance wheel, thus diminishing its inertia. The movement was given a free escapement, combining a wheel shaped like an eight-branched star and a pivoted detent.

With his two watches, named "watch A" (for *Ancienne*, or "Old One") and "watch S" (for *Seconde*, or "Second One"), made later, Pierre Le Roy established the principle of modern chronometry: free (or detached) escapement, balance isochronism, and thermal compensation of the balance. His success, however, did not give him the advantages obtained by Ferdinand Berthoud, his competitor. Discouraged, he abandoned his research.

For the 1795 prize, the Académie proposed the following subject: "Building a pocket watch appropriate for determining longitudes at sea, including divisions to indicate the decimal fractions of the day, to wit: the tenths, the thousandths, the hundred thousandths, in other words, dividing the day into ten hours, the hour into one hundred minutes and the minutes into one hundred seconds." The Académie Royale des Sciences being suppressed in 1793, the prize was not awarded. But the same subject was again proposed by the Institut National des Sciences et des Arts, in year IV of the Republic for the year VI prize. This prize was awarded to a nephew of Ferdinand Berthoud, Louis, who presented two watches, with the mottoes "Liberty Makes My Constancy" and "To Time, the Teacher."

While Pierre Le Roy and Ferdinand and Louis Berthoud were pursuing their work in France, two English watchmakers, John Arnold (1736–1799) and Thomas Earnshaw (1749–1829), separately but almost at the same time, were perfecting the type of chronometer that ultimately triumphed and was to be manufactured up to our day. They themselves took advantage of their models and produced about a thousand chronometers for a relatively low price.

The first of Arnold's chronometers, made in 1770, was a failure. In the following years Arnold improved his model and started a line of production. He gave it a pivoted detent escapement, a bimetallic balance to compensate for temperature changes, and a helicoid spring spiral. In 1780 he perfected it again by equipping it with a spring detent escapement.

Earnshaw, also in 1780, worked out the same type of escapement. Priority for the invention was bitterly contested by the watchmakers, as is shown in Earnshaw's book, published in 1808, *Longitude, an Appeal to the Public*. Like Arnold, Earnshaw commercialized his model of chronometer, whose escapement proved, in the long run, better than Arnold's. His spirals, meant to regulate the balance, were not however as perfect as his rival's.

Thanks to both watchmakers, the problem of calculating longitude at sea was definitely solved by the end of the eighteenth century. Through their reasonably priced chronometers, they brought within the reach of navigators the precise determination of longitude. Their work had thus crowned the laborious researches of John Harrison and Pierre Le Roy.

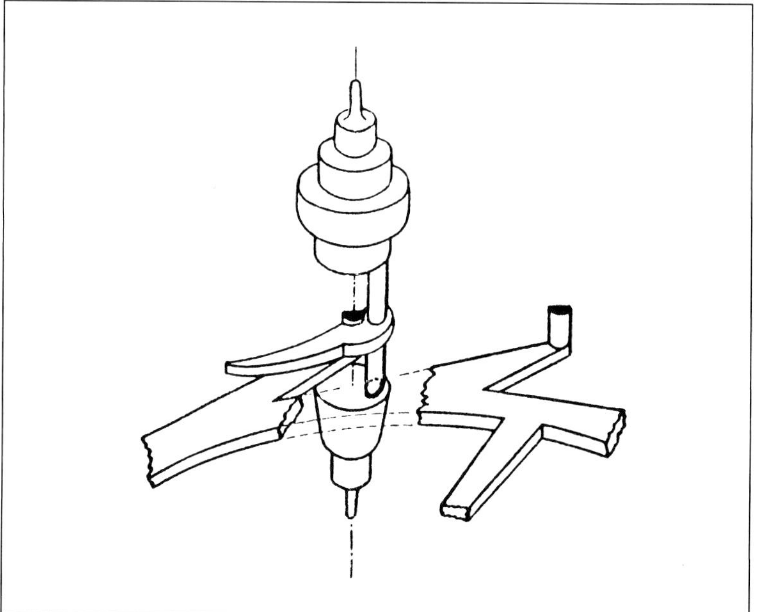

53 Virgule escapement.

54 Movement signed "Lépine Hger du Roy A Paris N° 5407." Around 1788. D. 50; Th. 10.
The movement, with bridges, has a virgule escapement and "wolf-teeth" wheels, characteristic of the work of Jean-Antoine Lépine (1720–1814), as is the dial, which presents both Arabic and Roman numerals.
London, Guildhall Library. The Clock Room, inv. 523.

55 Gold pair case watch. Movement signed "Thos Mudge London." Third quarter of the eighteenth century. Case dated 1759. D. 54; Th. 17.
The high repute of this watch rests on the fact that it is equipped with the first lever escapement invented by Thomas Mudge. Moreover, it has a micrometric regulating rack with a graded indicator shaped like a drum; thermal compensation through a bimetallic blade; a three-armed brass balance equipped with two flat spirals. A four-armed balance bridge has replaced the traditional cock.
Windsor Castle, H.M. the Queen of England Coll.

New Escapements

IMPROVEMENT OF THE CYLINDER ESCAPEMENT

The problem of metals to be used for the fabrication of the cylinder escapement was difficult to solve. The latten-brass wheel was replaced by a more resistant steel wheel by Thomas Mudge, Thomas Earnshaw and Abraham-Louis Breguet. To avoid the wear on the cylinder caused by the friction of the teeth of the wheel, Breguet and the great English watchmakers such as John Arnold used a ruby cylinder.

VIRGULE ESCAPEMENT

A close relative of the cylinder escapement, the virgule escapement was often used and perfected by Jean-Antoine Lépine (1720–1814), a brother-in-law of Caron de Beaumarchais, the inventor of the double virgule escapement.

The virgule escapement is composed of a wheel with vertical prismatic teeth and a curved part, similar to a comma (in French, *virgule*) fixed on the arbor of the balance wheel. The entry lift acts against the slanting face of a tooth. Then the tooth drops into a semicylindrical notch, where it rests during the back and forth motion of the balance. After disengaging, the tooth follows the concave face of the virgule; the action produces the second lift. After that lift, the next tooth comes to rest against the cylindrical surface preceding the hollow half-cylinder.

The difficulty of keeping oil in the rubbing parts and the fragility of virgule-bearing arbor explain the abandoning of this escapement around the end of the eighteenth century to the benefit of the cylinder escapement.

ANCHOR ESCAPEMENTS

Anchor escapements appeared in the eighteenth century but were perfected only in the next.

The first free, or detached, anchor escapement for watches was created by Thomas Mudge, who used it in a watch for George III, now kept in Windsor Castle.[34] The regulation of that watch proved very difficult, and Mudge gave up making watches with this escapement.

In the last quarter of the eighteenth century, many watchmakers undertook research on free escapements. Jean-Marie Pouzait (1743–1793), a Genevan watchmaker, presented a watch equipped with an anchor escapement to the Société des Arts in Geneva in 1787. Another Genevan watchmaker, Antoine Tavan (1749–1837), proposed a "crawfish-claw" escapement in 1796, then in 1805 a series of twelve escapements systems, including anchor escapements. Robert Robin (1742–1809), a watchmaker

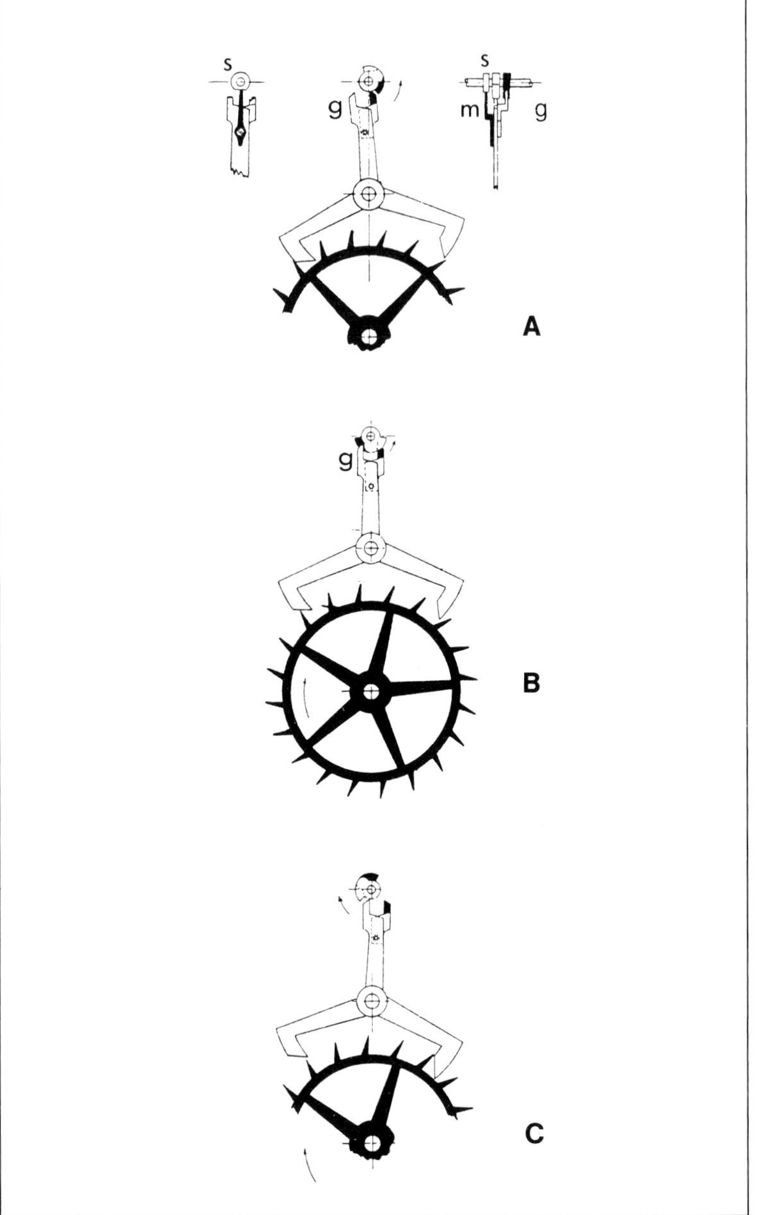

56 Lever escapement invented by Thomas Mudge (1715–1794) in 1757.
"The 3 main positions of the anchor [lever] can be observed from *A* to *C*. In *A*, a tooth is in contact with the rest surface of the entry claw (left). The balance wheel, freed from the anchor oscillates toward the left. When the safety dart *m* (in *A*) moves in front of the safety plateau, the black segment of the anchor fork is pulled toward the right (in *B*). The tip of the scape wheel tooth then is displaced under the rest surface of the left pallet of the anchor, raising it. Consequently, the top white horn of the fork *g* comes in contact with the top segment on the balance staff and transmits to it an energy-conserving impulse. Once the tooth has left the left side, the following tooth reaches the rest surface of the exit claw (right). The balance can then freely oscillate on its terminal arc (in *C*). During the recoil oscillation, the same phenomena are repeated, but in the opposite direction."
(R. Meis, *Les Montres de poche,* Fribourg, 1980, p. 19)

57a,b. Round watch. Gold case. Movement signed "Pouzait à Genève." D. 65.
The movement is equipped with an anchor escapement characteristic of the one invented in 1791 by Moïse Pouzait, used in conjunction with a large four-armed balance and beating the second. The dial, with the seconds in the center, presents the dates and the days of the week. The watch is the only one known to bear the signature of this watchmaker. Private coll.

to King Louis XVI, presented a new escapement inspired by both the anchor and the detent escapements. This sought to eliminate the inconvenience of the anchor's disengaging after a jar, as a consequence of which the fork would stumble against the plane of the balance. It was really perfected around 1830 by Georges Leschot (1800–1884), who found the angle that best ensured the anchor's stability during the free oscillation of the balance. The advantages of that construction encouraged watchmakers to choose the anchor escapement for quality watches.

The Lépine caliber

The honor for having created the modern shape of the watch belongs to Jean-Antoine Lépine. Around 1765 this remarkable watchmaker imagined a new movement construction that permit-

58 Demonstration model of an escapement signed "Ant^e Tavan Fecit anno 1805."
This pinwheel escapement is part of an eleven-model series that was the subject of a report published in 1831 under the title *Description des échappements les plus usités en horlogerie.* In its function, it is closely related to the anchor free escapement.
Geneva, Musée de l'Horlogerie et de l'Emaillerie

ted a switch from the placement of the wheelwork between two plates separated by pillars and the balance above the top plate.[35] He substituted for the traditional watch frame a single plate on which the various components of the movement, including the balance, were supported by independent bridges. The movement built thus, combined with a rest escapement (cylinder or virgule), no longer had to have a fusee.

French watchmakers, anxious to satisfy the demand of their clientele for flat watches, readily welcomed Lépine's invention. The English, contrarily, showed themselves more reticent about adopting the new caliber; however, it eventually was universally adopted in the manufacture of the modern watch.

A.-L. Breguet was the watchmaker who best understood the possibilities offered by Lépine's caliber and who utilized it so well that he created a new aesthetic of the watch, approved everywhere.

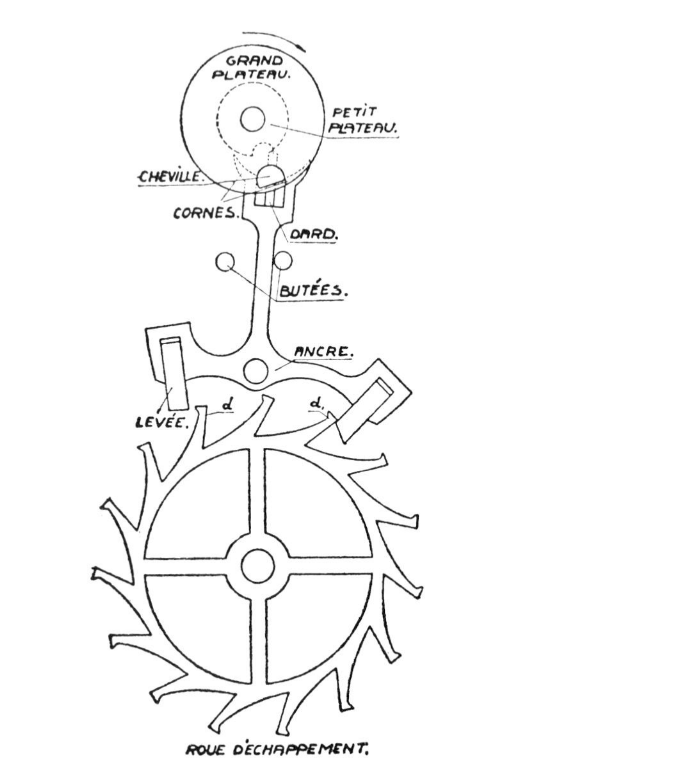

59 Anchor escapement (continental lever).

The inventions of Abraham-Louis Breguet

Through the boldness of his technique, his great number of inventions and the aesthetic and technical quality of his work, Breguet acquired an unparalleled renown, evidenced by the numerous studies and exhibitions devoted to him.[36].

His reputation was founded in the years following his installation in Paris in 1775 at no. 39, quai de l'Horloge, with the production of *perpétuelles* (self-winding) watches. The automatic winding system that characterized these watches was constituted by a plate-pivoted weight hanging on a spring, which wound up two springs contained in the barrel, by oscillating as the wearer

60 *A tact* round watch. Chased and enameled gold case. Movement ▷ signed "Breguet N° 2627." Paris, 1810. D. 40; Th. 7.5.
The watch, sold on October 22, 1810, for 600 francs to M. Titan, perfectly illustrates the extreme elegance of the style worked out by Abraham-Louis Breguet. The arrow-shaped hand, the eleven pearls arranged around the circumference (of which the three largest indicate the hours three, six, and nine) and the pendant marking twelve o'clock, allowed to know the time by touch. The movement has a ruby cylinder escapement with a *parachute* system.
Rockford (Illinois), Time Museum, inv. 577.

61 Round watch. Silver guilloche case and dial. View of the movement signed "Breguet n° 1187." Paris, 1810. D. 65.
The movement has the so-called natural escapement developed by Breguet, a *tourbillon* (with a four-minute rotation time), a compensated balance and a regulating spiral spring with a terminal curve. Jerusalem, L. A. Mayer Memorial, Institute for Islamic Art, inv. WA 108–71 (now stolen)

walked. The weight was locked when the springs were sufficiently wound. On the dial side, a gauge seen through a window indicated the running reserve of the movement, which was sixty hours, under the condition that the winding mechanism would function for half an hour.

Breguet's forced stay in Switzerland during the French Revolution, gave him the leisure to develop many inventions that he was later to commercialize successfully. Thus, during the period, he perfected Graham's cylinder escapement by using a tile-shaped cylinder made in ruby, which was adjusted below the bottom pivot of the balance. The escape wheel and the cylinder were thus located between the dial and the plate. The balance pivot, rolling in a hole dug into a hard stone, had the advantage of being stronger than other cylinder constructions, needing no indentation to lodge the cylinder.

The problem posed by variations in running, due to changes in position held Breguet's attention entirely. The solution he worked out shows his capacity for invention. He decided not to attach the balance and the escapement to the plate, but to mount them on a mobile frame that completed a revolution every six minutes; he called them *tourbillons*. The precision brought about by this arrangement, for which Breguet received a patent in 1801, ensured a large success to the *tourbillons*. The best watchmakers in the nineteenth century built some of them, such as Jacques-Frédéric Houriet, Urban Jürgensen, Frédéric-Louis Favre-Bulle.

From the beginning, watchmakers had created dials allowing one to tell time by touch, placing studs where the hand indicated the hours. In the eighteenth century the repeater was sometimes accompanied by this kind of mechanism permitting one to tell the time in darkness or whenever one wished.

More discreet than ringing of the *au passage*, the method was nevertheless somewhat noisy. Breguet understood that a model

62a,b. *Subscription* watch. Chased silver case. Movement signed "Breguet n° 3492." Paris, 1820. D. 62; Th. 16.
The movement has a ruby cylinder escapement and a *parachute* system. The one-handed dial is in white enamel. This watch was sold for 611 francs on September 11, 1820, to Count Demidoff.
La Chaux-de-Fonds, Musée International d'Horlogerie, inv. 357

indicating the time in darkness through a simple and discreet system, using only the sense of touch, would be welcome by his sophisticated clientele. The watch he conceived presents on the face of the case a long, sturdy hand meant to mark the hours, which were indicated by twelve studs on the rim.

Many other innovations by Breguet improved watchmaking technique. Thus, his *parachute* system was perhaps the first to prevent damage to the balance arbor through jolting. The hard stones in which the pivots rolled were mounted on the arms of steel springs that were elastic enough to absorb the shocks.

Breguet's return to Paris in 1795 marked the beginning of the most brilliant period of his career. The extreme originality of his work and the perfection of his execution won him an international renown that only can be compared to that of the greatest artists. His workshop, which included the best craftsmen in France, was then supplying the whole aristocratic and financial elite in the world with clocks and watches. His watch production included "simple watches," that is, without complications, "perpetuals" (self-winding watches), *tourbillons*, "subscription watches," repeaters and watches with multiple complications.

Subscription watches make up an interesting chapter in the history of watchmaking, a measure of Breguet's enterpreneurial spirit and talents. The manufacturing of these watches started in

63

64a *64b*

◁ 63 Pocket marine watch. Engraved silver dated 1776 (N crowned). Movement signed "Les frères Goyffon A Paris 837." D. 63; Th. 32.
This watch is of fundamental interest because of analogies with Pierre Le Roy's work. It has a free escapement built on the model of the chronometer by Pierre Le Roy that is kept in the Musée National des Techniques (ill. 50). Bimetallic thermal compensation acts on the length of the spiral, in a way that is similar to that of the device used by Pierre Le Roy for the "small round one."
Paris, Musée National des Techniques (CNAM), inv. 21962

64 a,b. Double-dial round watch. Engraved gold case. Movement by Abraham-Louis Breguet bearing the number *92*. Paris, 1805. D. 60; Th. 22.
This watch has a ten-minute repeater. On the face (a) covered by white enamel dial, the date, the days of the week, the equation and the individual seconds can be seen. On the other face (b), covered by a chased gold dial, the phases and the age of the moon appear. The watch was sold on thermidor 11, Year XIII, of the Republic (July 30, 1806), to the duke of Praslin for 4,800 francs.
Paris, Musée National des Techniques (CNAM), inv. 16311

the months following his return to France. The customers "subscribed" for a watch by paying a deposit with their order and the balance on delivery. The procedure allowed Breguet to have investment capital at his command to buy machinery for the production of his watches. With their high technical quality and understated elegance, the subscription watches also had the advantage of being sold at an affordable price. Their success encouraged Breguet to keep up the line until his death in 1823. At least fifteen hundred watches of this type are estimated to have come out of his shop in his lifetime. With a fairly large diameter, between sixty and sixty-two millimeters, they had a ruby tile cylinder escapement and a train limited to four wheels. Breguet had indeed decided to do without the minute hand, probably with a view to reducing costs. The slender end of the lone hand permitted one to read the time to the minute. More than any others, these watches spread the simple, even austere, Breguet style, which announced the functional aesthetic of the industrial age.

Among the prestigious watches in which Breguet piles up the complications he invented or perfected, the most famous is the one known under the name of "Marie-Antoinette watch," kept, before its theft, in the L. A. Mayer Collection of the Institute of Islamic Art in Jerusalem.[37] Ordered in 1783 and finished forty years later, the watch has, besides automatic winding, which classifies it among the "perpetuals," a perpetual calendar, a thermometer and a repeater striking on the hour, the quarter hour and the minute.

All the watches that came out of the Breguet workshop, even the subscription watches, were luxury creations meant for a limited clientele. The average-quality watches, as well as jewelers' watches of the end of the eighteenth century and the start of the nineteenth, obviously did not receive all the improvements brought to horological technique by Breguet, Lépine or Arnold. Nevertheless, under the influence of the "original creations" of these watchmakers, and more often thanks to improvements discovered in the first half of the eighteenth century, movements were little by little transformed.

65 Round watch. Painted enamel gold case. Movement signed "Hessen A Paris." About 1790. D. 55; Th. 14.
Equipped with a cylinder escapement and a barrel sticking out above the plate, the movement presents a flat caliber characterizing the production of André Hessen, master in 1775, watchmaker of Monsieur, the king's brother. He described this construction in his *Mémoire sur l'horlogerie* (London, 1785): "I shape on the balance plate an opening through which the barrel passes in such a way that the height in thickness (of the frame) is equal to that of the pillars" (p. 23).
Paris, Louvre, Olivier Coll., inv. OA 8593

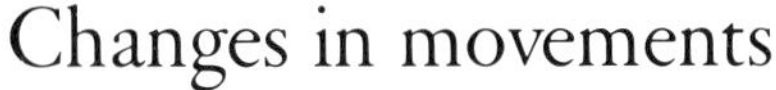

Changes in movements

French and Swiss watches

Most French and Swiss watchmakers of the end of the eighteenth century continued to produce watches equipped with a verge escapement. Influenced by the vogue of flat watches, they considerably reduced the frame of the movement. In 1785, in an essay on horology published in London, André Hessen underlined the faults of flat watches with verge wheel escapement: "To the ostentatious charm of a very complicated watch, they wanted to unite the advantage of smallness and the flat shape. The Artist has been forced to yield to the taste of the day, but has beauty of skillful execution been accompanied by an equal perfection of the watchmaking science? No, assuredly, since up to the present he has only been working on principles that do not yet appear invariably determined, and he has had to injure those that are recognized, just to impose on them the stamp of fashion, & finally elegance of shape has been obtained only at the expense of solidity." Hessen proposed the construction of a flat watch that had the advantage of not harming the running of the movement. The caliber he had envisioned kept the traditional arrangement of a movement between two plates, but the top plate was pierced at the location of the barrel so that it would be of a suitable height. The plates were open at the site of the crown wheel, in order to use an escapement with reasonable dimensions.

Every high-quality watch made in Louis XVI's reign had a cock-head in which the top pivot of the balance arbor rolled. The use of pierced gemstones was still very rare. Because of Julien Le Roy, the adjustable potence became generally used in the French as much as in the Swiss workshops.

English watches

Until about 1800 most English watches were built with a pair case. As on the Continent, their movements tended to become smaller.

English watchmakers often used the cylinder escapement. On their quality watches they usually employed the duplex escapement, invented by J.-B. Dutertre and later perfected by Pierre Le Roy. The anchor escapement was not very frequently used. Before 1800 only a few watchmakers, such as Josiah Emery, John Grant, John Leroux and Richard Pendleton, built watches equipped with that escapement.

66a

66b

67

68

IV The triumph of precision in the nineteenth century

In the area of the production of high-precision watches, the nineteenth century and the beginning of the twentieth was an auspicious period. Watchmakers and scientists united to solve problems brought to light by their predecessors—linked to the escapement, the isochronism of the balance and thermic compensation.

66 a,b. Round watch. Chased gold case. Movement signed "Urban Jürgensens Sönner Rishenhawn." Denmark, about 1830. D. 60; Th. 22. This *tourbillon* watch was made by Louis-Urban Jürgensen (1806–1867), son of the famous Danish watchmaker Urban Jürgensen (1776-1830). It has a spring detent escapement, a spherical spiral and a two-armed balance with prismatic masses. The influence of the Locle watchmaker Jacques-Frédéric Houriet—whose daughter married Urban Jürgensen—is clear in its construction. In the bottom part of the dial, there is a thermometer.
La Chaux-de-Fonds, Musée International d'Horlogerie, inv. 496

67 Lady's watch. Enameled gold case matched with a brooch. Movement by Patek Co., no. 4536. Geneva, 1851. D. 29.5.
Diamond roses form a floral motive in the center of the bottom, which is covered by an opaque blue enamel. The movement, equipped with a cylinder escapement, has a remontoir. The watch was sold on August 15, 1851.

68 Pocket marine watch. Movement signed "Par L^s Berthoud N° 2402." 1791. Rear of silver case. D. 60; Th. 23.
This chronometer, which carries the number 18 engraved on the bezel mount, is described under "number 2402" in the sales book kept by Louis Berthoud (Musée National des Techniques, archives 8° 189). On September 18, 1791, it was sold to the chevalier d' Entrecasteaux. Its movement includes a pivoted detent escapement, a bimetallic three-armed compensated balance and a helicoid spiral.
Paris, Musée National des Techniques (CNAM), inv. 19255

The improvements brought by Abraham-Louis Breguet and Georges Leschot to the anchor escapement were applied by all the great watchmakers of the century. As we already underlined, they also used the *tourbillon* escapement in their most beautiful creations.

The isochronism of the balance, whose quality determined the regulation of watches, was a constant object of research for watchmakers like F. Berthoud, A.-L. Breguet and J. Winnerl. The problem was finally solved by the French scientist Edouard Phillips, a member of the Institut. In 1861 he applied to the regulating spiral spring the mathematical formulas he had discovered during his experiments on spring elasticity and thus established the principles of the theory of forms to be given to the terminal curves of spirals.

In the regulation of precision watches, the quality of the thermal compensation was also very important. During the second half of the nineteenth century, palladium was used with success for manufacturing spirals and balances. In the beginning of the following century, Guillaume, the director of the International Bureau of Weights and Measures in Paris, discovered an alloy, practically unchanging at various temperatures, the *élinvar* which drew its name from the contractions of the words *élasticité invariable* (invariable elasticity).

The elimination of the key, replaced by the remontoir, marked an important practical advance. Adrien Philippe presented his first remontoir watches in 1842; they did not receive the favor of the watchmakers. However, two years later, during the Industrial Exhibition held in Paris, the invention attracted the Genevan watchmaker Patek, who offered him a partnership that turned out to be very fruitful. The remontoir watches became one of the most famous specialties of the concern thus founded under the name Patek & Philippe.

In the second half of the nineteenth century, precision watches, whether simple or with complications, won remarkable results in regulation tests performed in observatories in Besançon, Neuchâtel and Geneva. The Swiss watchmakers excelled in that type of production. In his report on horology presented to the World Exposition of Paris in 1878, Claudius Saunier remarked: "Switzerland is unrivaled for complicated watches. In this line, it produces admirable works."[38]

69 Round calendar watch. Gold case. Inscription on the cuvette: "Créée et exécutée par Pierret à Paris 1855 JAD 16534." D. 55; Th. 10. This seconds watch, in which windows on the face reveal the day, the date and the month, is an example of the research done by watchmakers to improve their art. Pierret described his construction in a work published in 1885, *Horlogerie, outillage et mécanique.*"
Paris, Musée National des Techniques, inv. 8336

Part three

Decorative evolution

I Decoration from 1500 to 1675: international fashions

The golden age of watch decoration is situated at the time when horologic science was still in its infancy. Mediocre timepieces – the sixteenth-century watches and those of the first three-quarters of the seventeenth – were true jewelry. Every decorative technique, the most refined and the most sumptuous, were applied to them. Goldsmiths, lapidaries, jewelers, engravers and enamelers, encouraged by the social elite for whom their watches were intended, created cases of varied shapes, whose decorative perfection remains unequaled.

The multiplicity of fashions, whether ephemeral or enduring, that affected the ornamentation of watches during the period, proves the interest of the craftsmen and their rich patrons for these objects of attire, worn attached by a chain around the neck or the waist.

French, German, Genevan and English watchmakers were then signing their names on watches reflecting the same fashions. The continuous streams of horological exchange centers, the influence of French creations, spread particularly by collections of engravings or by many immigrating French artists, largely contributed in giving an international style to the decoration of watchcases.

The first watches (c. 1500–1580)

When the first portable clocks were manufactured, two forms triumphed: watches shaped like small drums and spherical watches.

Drum watches

The drum watches that have reached us are relatively numerous, for they were not only in fashion for nearly a century, but at the time of their manufacture, they were already the most common type of watches.

Because of the modesty of their material, which did not tempt the greed of men, the cases of drum watches in gilt brass are still extant. But other, more precious metals were used in the manufac-

70 Drum watch. Engraved and pierced gilt brass case. Anonymous movement. Germany, sixteenth century. D. 56; Th. 23.
In the center, the dial lid is ornamented with a star and foliage. Around its circumference, heart-shaped openings let one see the hours in Roman numerals (I to XII). When the lid is open, another hour appears, in Arabic numerals (13 to 24), and small studs let the time be known by touch. The movement has an *au passage* striking train.
Paris, Louvre, Olivier Coll., OA 8398

◁ 71 Engraved gilt brass watch shaped like a drum. Germany, sixteenth century. D. 59.6; Th. 24.
The dial presents a double numbering, both in Roman and Arabic numerals. Openwork lid. Its inside face is engraved with leaves and heads.
London, British Museum, inv. 88, 12-1, 152

72 Spherical watch. Engraved and pierced gilt brass case. Movement signed and dated "Jacques Delagarde Bloys 1551." D. 55.
The case is engraved with interlacings, foliage and grotesque masks. The top hemisphere, carrying the suspension ring, is pierced so as to let the striking of the movement be heard. The dial is slightly sunk into the bottom part. The iron and brass movement has an annular foliot and a tall fusee equipped with a gut cord.
A king's horologist, Jacques de La Garde was active in Blois from 1540 to 1580. He had several sons, three of whom became watchmakers.
Paris, Louvre, Garnier Coll. OA 7019

ture of primitive watches. In the inventory of Florimond Robertet's possessions, drawn by his widow in 1534, figure "twelve watches, of which seven are ringing and the other mute in gold, silver and brass cases. . . ."[1] The pieces that have survived let us imagine the jewelry masterpieces then created to enclose the movements.[2]

The drum-watch type clearly derived from the small cylindrical table clock. The dial, first unprotected like a table clock's, was covered with a full lid and later, pierced over each numeral to permit the reading of the hour. Both the top and bottom plates close the cylindrical body of the case, their molded ridges slightly overlapping its circumference. Each one opens on a hinge fixed on the level of XII.

The lid decoration often shows refinement. The windows that reveal the numerals have varied contours. The most prevalent are shaped like arcades, quadrilobes, hearts or pears. In the more ambitious compositions, the numerals can be seen among figures holding hands in a carol dance. The lid's center and the bottom and the periphery as well are decorated with foliage populated by fantastic animals, a scene or a portrait.

German cases introduce molten and cast decoration, whose high relief is tinged with a certain heaviness, contrasting with the delicacy of the engraved and chased decoration of the French cases.

The composition of dials is not very varied. The watchmakers of the sixteenth century were doubtless conscious of having established a very readable type of dial. On the circumference, the Roman numerals I to XII, indicate the hours until noon. Between these, a design, such as a fleuron, a losange, a star or an arrow, marks the half hours. A second circle, numbered in Arabic numerals from 13 to 24, indicates the hours until midnight. In the German pieces, one notices the characteristic Z shape of the 2.

When the movement includes an alarm, a moving circle, numbered 1 to 12, lie in the center of the dial. The center is very often engraved with a sun whose rays, directed toward the hours and the half hours, punctuate the movement of the hand. The latter is long, in blued steel; to avoid confusion, the extremity marking the hour is in the shape of a fleuron or a point, while the other end is shaped with a circular ornament. Jutting studs, located above the Roman numerals, let one know the hour by touch in the dark.

Most drum watches still extant are German. They sometimes are dated and marked with the initials of the watchmaker. At times, a hallmark indicates the place of manufacture. The oldest watch belonging to the Uhren Museum in Wuppertal bears on the inside of the bottom the date 1548 and the letters *CW*, separated by a head. These could be the initials of a watchmaker active in Nuremberg from 1527 to 1557, Caspar Werner.[3]

Spherical watches

The spherical watches borrowed their shape from the "civets," also called "musk apples," mostly used in Venice and the East; these were small chased metal spheres, pierced with many holes to let out the fragrance of the perfume they contained.[4]

Contemporary evidence and a few rare pieces that have reached us prove that such watches were built in all watchmaking centers of the time, such as Nuremberg, Blois and Paris. Johann Neudorfer, in his *Accounts of the Main Artists and Craftsmen Who Lived in Nuremberg over the Past Hundred Years*, published in 1546, points out that "Henlein is almost one of the first to have invented the building of small watches in civets."[5]

The oldest French watch, kept in the P. Garnier Collection in the Louvre (inv. OA 7019), is spherical. It carries the signature of a Blesian watchmaker, Jacques de La Garde, and the date 1551. A German-made watch, noted in the Ashmolean Museum in Oxford, is closely related.[6] With an almost identical shape, these two watches present a decoration of similar interlacings. The German case was cast in a mold, while the French case was delicately chased. The techniques of the German and French goldsmiths were different, but the ornaments and the forms they had chosen were the same.

The case of spherical watches is composed of two equal parts opening with a hinge and closing with a small movable rod, locked with a catch. It is often decorated with a symmetrical composition of widening or narrowing galoons that intersect, forming knots or arching over a background of delicate foliage. That interlacing decoration of Eastern inspiration, so-called moorish tracery, is also found in bookbinding, goldsmiths' work and ceramic ornaments of the second quarter of the sixteenth century.[7] The dial, similar to that of a drum watch, is lodged at the base of one of the hemispheres. The base rests on a ring molded with several ressauts, which forms the foot. The pendant is fixed at the vertical of the dial.

The watches lodged in "musk apples" give the first concrete evidence of the refined fantasy and the originality of the creations of small-work horology in the sixteenth century. They prefigure the rage of watches affecting shapes of every kind of flower or animal.

Oval and octagonal watches (1570–1640)

Oval or octagonal elongated watches were in great favor from the last quarter of the sixteenth[8] until the beginning of the following century. Museums have collected a fairly large number that testify,

73, 74, 75, 76 Oval watch. Engraved silver and gilt brass. Movement signed "P Cuper A Bloys." Around 1600. L. 40; W. 30.
The lid of the movement represents Narcissus falling in love with himself while admiring himself in a fountain; that of the dial has Orpheus charming the animals with the sweet accents of his music. Both engravings are after two Delaulne etchings (Bibliothèque Nationale, Cabinet des Estampes, Ed 4 pet. fol.).
Paris, Louvre, Garnier Coll., inv. OA 7022

73

74

75

76

77

78

77, 78 Oval watch. Engraved silver and gilt brass case. Movement signed "J Barberet A Paris." Around 1600. L. 51; W. 41.
The lid of the movement represents a warrior figure (Mars?) after a Delaulne etching (Bibliothèque Nationale, Cabinet des Estampes, Ed 4 pet. fol.).
Paris, Louvre, Garnier Coll., inv. OA 7028

through the variety of signatures they carry, to their international vogue.

The cases are generally made of brass or silver. The two metals are at times used jointly to obtain a decorative effect. Often, in this case, the silver lids and side surface contrast with the gilt brass nervures. The dimensions of these cases vary from 4 to 5 centimeters in length, from 2.5 to 3.5 centimeters in width, and from 2 to 3 centimeters in thickness. The lid of the dial and the one that covers the movement plate are slightly convex, while the periphery is straight. The lids are generally held by two hinges on either side of the pendant.

Engravers gave those watches a style characterized by mannerism, delicacy and richness in the decoration. Surrounded with flowery arabesques, landscapes, allegorical figures, and mythological or religious scenes adorn the outside surfaces of the cases and the dial. The inside of the lid is often engraved with a wreath of foliage, at times framing a monogram, a motto or armorials. The pendant almost always has a crest or a bud-shaped base; the bow passing through its head, is a ring on which the watch hung. Oval watches usually had a molded stud placed opposite the pendant, which helped enhance the harmony of the shape of the case. The gilt brass dial presents a silver horary circle applied in relief. Between the numerals, a design indicates the half hour. Dials from the second quarter of the century sometimes have a circle marking the quarter hours.

79, 80 Oval watch. Engraved silver case. Movement signed "Aug Buschman." Augsburg, end of the sixteenth century. L. 84; W. 51; Th. 40. The bottom of the case is engraved after an etching by Delaulne, representing two vestals next to an altar (Bibliothèque Nationale, Cabinet des Estampes, Ed 4 pet. fol.).
La Chaux-de-Fonds, Musée International d'Horlogerie, inv. 1147

Whether signed by French, Swiss, German or Dutch watchmakers, these watches reveal a stylistic unity explained by the fact that their decorations were copied or inspired from the same etchings, contained in collections printed by mannerist *petit maîtres* of the second half of the sixteenth century, such as Etienne Delaulne, Théodore de Bry and Solis the Younger, or by ornamentists of the first half of the seventeenth century, such as Antoine Jacquard, Jacques Hurtu and Michel Le Blon.

Etienne Delaulne, who was born in Orléans around 1518 and died in Strasbourg in 1595, was a goldsmith and engraver who published many models of oval watches.[9] He offered engravers mythological scenes or isolated allegorical figures intermingling with grotesque decorations or standing out from a naturalist landscape or from foliage scrollwork.

Several oval watches in the Garnier Collection in the Louvre offer engraved decorations copied from Delaulne's etchings. One (inv. OA 7022), signed by P. Cuper, the Blesian watchmaker, is decorated inside the bottom cap with *Narcissus Falling in Love with Himself* and on the top lid with a scene representing *Orpheus Charming the Animals with the Sweet Accents of his Music*. Another one, with a movement by J. Barberet of Paris, has the figure of a warrior surrounded by grotesques (inv. OA 7028). An engraving representing a birdcatcher laying his snares was used as the model for D. Martinot of Paris. A watch kept in the Musée International d'Horlogerie (inv. 1147), signed Buschman, presents a cap bottom engraved after an etching by Delaulne, figuring two vestals watching the sacred fire. The engraver simplified his model, retaining only the essential elements and substituting for the decoration of grotesques a simple background of strawberry flowers and foliage.[10]

Antoine Jacquard, whose career as a goldsmith and harquebus maker was spent in Poitiers in the first half of the seventeenth century, published many models of dials and cases[11] influenced by Delaulne. Thoughtfully conceived in order to be copied without modification, his engravings stage mythological scenes like *The Rape of Ganymede, Diana's Bath, The Judgment of Paris, The Fall of Icarus* and *Aeneas Carrying Anchises*. When the scene occupies the whole surface of the bottom cap, the figures stand on a frieze resembling a platform, under which an allegorical figure–for instance a river god–is represented or there is foliage scrollwork. If the scene is framed in a medallion, particularly on the octagonol cases, whose circumference is made up of panels, it is surrounded by flowery arabesques, interrupted on the top and the bottom by couples of cupids, children tasting fruit, chained slaves, lovers or horsemen.

Jacquard's plates meant for dial decorations propose similar ornamental compositions. The scene occupies the surface delimited by the horary circle, while foliage scrollwork with figures decorates the circumference. Jacquard also provides frieze models to be engraved on the sides of the oval cases: small mythological scenes, cupids, tritons and allegorical figures such as Time interrupting the interplay of graceful arabesques.

An oval watch in the P. Garnier Collection, signed by P. Combret of Lyons[12] (inv. 7039), has a decoration probably copied from the Jacquard's models. The engraved topics, presented above a frieze, are among those that inspired the engraver several times: *Acteon Changed into a Stag* and *Diana Discovering Callisto's Condition*. The dial is similar to those that Jacquard proposed: within the hour circle, a scene figuring Pyramus and Thisbe; above the XII, on the periphery, ornamented with flowery foliage, two cupids holding a basket of fruit, and immediately below the VI, a child watching a she-bear drinking from a fountain. The frieze that unfolds on the side is composed of foliage, punctuated by flowers, surrounding the figure of Minerva, as well as a scene copied from a Jacquard model, *Leda and the Swan*.[13]

Michel Le Blon (1587–1656), a goldsmith and engraver and a student of Théodore de Bry, worked in London and then in Amsterdam, where he died. He published models of oval watches, whose decoration is essentially composed of interlacing foliage populated by animals and ending in flat-petaled flowers. His creations undoubtedly influenced the engraver of the watch signed around 1620–1630 by the London watchmaker Jo. Midnall, kept in the Pierpont Morgan Collection (Metropolitan Museum, New York), and the one by N. Rigdale, belonging to the Victoria and Albert Museum (inv. 2365–1855).

This international style, dominated by French fashions, which were propagated by prints, can also be explained by the orders for cases that watchmakers sent from everywhere in Europe to the Blesian goldsmiths, whose renown was then considerable.

A watch in the Guildhall Museum in London has a star-shaped case, whose author inscribed his name on the dial: "de Heck Sculp." He is the engraver Gérard de Heck, established in Blois in the first third of the seventeenth century. The movement of this watch carries the signature of David Ramsay, the famous watchmaker of King James I. A watch that belonged to this king, exhibited in the Victoria and Albert Museum, is also signed by Gérard de Heck and David Ramsay. The sovereign's portrait, engraved inside the lid, was executed from an engraving by Simon van de Passe, printed in 1618. The signatures on both these watches give the formal proof that an eminent English watchmaker ordered cases, possibly routinely, from an engraver established in Blois.

More evidence of the influence of French creations is provided by an oval watch in the Victoria and Albert Museum (inv. 921–1855) dating from the first quarter of the seventeenth century and signed by a Worcester watchmaker. It has a case engraved with a biblical episode whose French inscription, "*Daniel Chapitre 3*," proves that it was either imported from France or copied in England from a French model.

As authorized by their Community statutes, a few watchmakers practiced the art of engraving. But most called on the service of engravers. Thanks to Abbé Develle's research,[14] a list has been made of the most important metal-engraving shops active in Blois in the first half of the seventeenth century. Michel Vauquer gave up the watchmaking profession to devote himself entirely to engraving until his death in 1630. The shop that Abel Bérault II opened around 1623 was important for more than fifty years. Christophe Morlière's business, engraving and enameling, also prospered until his death around 1644. No doubt attracted by the success of the Blesian masters, numerous foreign engravers, chiefly from Geneva, Flanders and Holland, settled in Blois. Among them, Robert Le Roy came from Rotterdam before 1605, and Gérard de Heck married in his adoptive city in 1611. In Blois, engravers and watchmakers formed a real artistic community, mainly devoted to the creation of watches, that imposed its style of work on all the other horological centers for a long time.

The oval and octagonal watches decorated with delicate engravings are, for art history, precious evidence of goldsmiths' work in the end of the sixteenth and the first half of the seventeenth century.

81

82

81 *Left.* Oval watch. Engraved silver and gilt brass. Movement signed "N Rigdale." London, second quarter of the seventeenth century.
The bottom and the lid of the case offer a similar foliage decoration in the style of the ornamenter Michel Le Blon.
Right. Oval watch. Silver and gilt brass case. Movement signed "Ri Barnes at Worcest." England. Early seventeenth century.
The lid, with the reference Daniel 3, and the case bottom, marked "Daniel Capt," are decorated with scenes from the reign of Nebuchadnezzar.
London, Victoria and Albert Museum, inv. 2365–1855 and 921–1855

82 Star-shaped watch. Silver case signed by Gerard de Heck. Movement signed "David Ramsay Scotus me fecit." Blois, London, about 1625. L. 45.
The case, decorated with biblical scenes is the work of a Blesian artist, while the movement is by a renowned Scot watchmaker who, in 1632, was the first master of the Clockmakers' Company. This watch is evidence of the international exchanges existing in the watch industry in the seventeenth century.

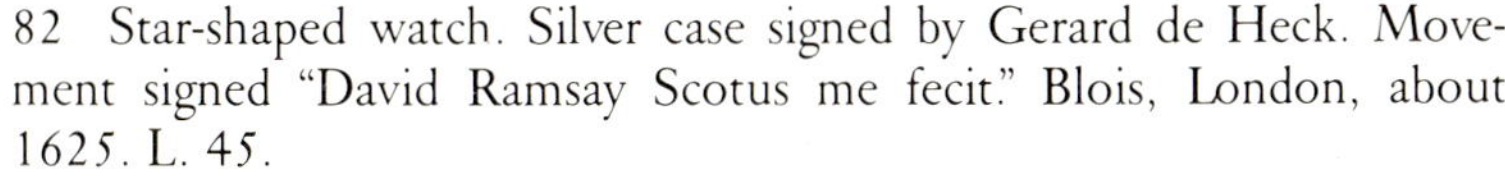

London. Guildhall Library, The Clock Room, inv. 7

Form watches (1570–1660)

The fashion of fancy watches, along with the one of oval watches, appeared in the last third of the sixteenth century at the latest and remained in favor until the mid-seventeenth century. Worn hanging from the neck or at the waist so as to enhance one's attire, these watches were real jewelry, assuming varied and refined shapes, often done in metal or in rock crystal and less often in other stones.

The cruciform watches with rock crystal cases were very much in fashion, if one can judge from the relatively numerous surviving specimens. Their shapes are many: simple latine cross or double-armed, flory cross, rounded-arms cross, Maltese cross.

The cuvette (cap) and the lid are cut into facets and framed with metal armatures supporting the hinges fixed above XII. These maintain, besides the lid and the bottom, the movement and the dial. Like most elongated watches of the time, the cruciforms have a fairly long molded stud, affixed opposite the pendant. This stud is at the same place as the pearl hanging from the lower part of the pendentives in vogue during the sixteenth century, enhancing in a similar way the harmonious shape of the case.

The brass or silver armatures are often engraved with scrollwork, leafy branches or geometric designs. The dial is usually silver, while the cruciform plate that supports it is made of brass.

From the first half of the seventeenth century, Geneva appears to have been a center specializing in fancy watches. Most of the rock-crystal cruciform watches that have been preserved indeed bear the name of a Genevan watchmaker: the Horology Museum in Geneva owns two of them signed by Jacques Sermand and Pierre Duhamel; the British Museum preserves one, whose author is Jean Rousseau (inv. 74, 7–18, 28); the Pierpont Morgan collection has one by Jean Cusin. A few are signed by French watchmakers, such as the one in the P. Garnier Collection made by the Lyonese, Jean Vallier (inv. OA 7040), or the one in the Dutuit Collection bearing the name of a watchmaker from La Rochelle, Fonneveaux (inv. 1468; Petit-Palais Museum, Paris).

Cruciforms were also made in metal. Some are goldsmiths' masterpieces whose decoration proceeds from religious inspiration and unites engraving, chasing and repoussé work. Made around 1630, the cruciform watch with the movement signed by O. Tinelly in Aix, kept in the British Museum (inv. 88–12, 1–187), is decorated with admirable repoussé plates, applied on the gilt silver lids of the case. The dial lid is decorated with *The Crucifixion*, the one of the movement with *The Virgin and Child*. The periphery, in silver, is engraved with *The Instruments of the Passion*.

The fashion of "skull-shaped" watches reflects the taste of the seventeenth century for memento mori. No other cases could better express the inexorable march of the timepiece's movement. Made in metal, the skulls were sometimes engraved with religious scenes accompanied by moral aphorisms. The skull watch of the Olivier Collection, signed Jean Rousseau, partly pierced on account of the striking movement, provides a remarkable example (inv. OA 8314). Most often, the metal skulls are hammered, worked in *repoussé* and polished; only the dial receives an engraved decoration. Their diversity rests in the more or less rounded shape of the skull and in the manner that the diverse parts of the case are assembled.

When the hinge is located behind the cranium, the lower jaw is used as the lid for the dial; such is the case with Jean Rousseau's silver watch kept in the Cluny Museum (inv. 18530). The hinge may, on the other hand, be placed under the chin as it is on the watch of the Garnier Collection signed by the Genevan watchmaker Duboule (inv. OA 7036). The memento mori watches made in rock crystal are more rare. A fine example is the one signed by Jacques Joly in the Olivier Collection (inv. OA 8288).[15]

The goldsmiths and lapidaries borrowed all kinds of forms from the vegetal and animal world.

Tulip bud watches were particularly appreciated by a clientele sensitive to the vogue of that flower which came from Holland around 1625. The petals, mounted on hinges, may be in rock crystal and set in metallic armatures. Jacques Sermand signed two watches of this type: one belongs to the Garnier Collection (inv. 7063), the other to the Metropolitan Museum of Art. Another watch of the kind, signed Duboule, is kept in the Cluny Museum (inv. 18515); the rock crystal of which its case was made, is of a remarkable transparency. The petals of tulip-watches may also be in metal with *repoussé* nervures, like those of a watch signed Henry Grendon,[16] or ornamented with enamel like those by J. Jolly belonging to the Garnier Collection (inv. OA 7032).

Fleur-de-lys cases were also made in metal or rock crystal. Three petals in polished silver, cover a movement build by a Nérac (Lot-et-Garonne) watchmaker, J. Dracques (Garnier Collection, inv. OA 7047).

83 Cruciform watch. Gilt silver case, decorated with repoussé scenes. Movement by O. Tinelly. Aix-en-Provence, around 1630. L. 63.5; W. 42.

This piece is a good example illustrating the fashion of metal or rock-crystal cruciforms in the late sixteenth century and the first half of the seventeenth.

London, British Museum, inv. 88–12, 1–187

84

85

86 Tulip-shaped watch. Rock-crystal case. Movement signed by Duboule. Geneva, first half of the seventeenth-century. L. 55.
Most watches, like table clocks, were delivered in a protective case. This is one of the very few which kept its case.
Château d'Ecouen, Musée National de la Renaissance, inv. Cl 18515

◁
84 Skull-shaped watch. Engraved and pierced silver case. Movement signed "Jean Rousseau." Geneva, middle of the seventeenth century. D. 55; Th. 45.
The cranium forms the cuvette, while the lower jaw, articulated by a hinge placed in the back of the head, is used as the dial lid. Two reserved medallions are engraved, one on the forehead figuring Adam and Eve, the other in the back of the skull, representing the Resurrection. Sayings from the Gospel surround the decoration.
Paris, Louvre, Olivier Coll., inv. OA 8314

◁
85 Skull-shaped watch. Brass case. Movement signed "Duboule." Geneva, first half of the seventeenth century. D. 32; Th. 28.
This memento mori watch reminded its owner of the flight of time not only through its shape but also through the decoration engraved on its lid, showing a sitting child holding a skull in one hand and an hourglass in the other.
Paris, Louvre, Garnier Coll., inv. OA 7036

The "shell-shaped" watches, most often made in rock crystal, were very much in favor. Their forms are varied: elongated shells with sharp angles, oval shells, shells called "combs of Venus," lobate shells. The Paul Garnier Collection offers diverse samples of them.

Animal-shaped watches seem to have been a Genevan specialty. A dog-shaped watch signed by J. Joly is preserved in the British Museum (inv. 88–12, 1–206). One, shaped like a rabbit, signed Pierre Duhamel, is kept at the Museum of Horology in Geneva. Another, shaped like a lion, bears Jean-Baptiste Duboule's signature.[17]

Cases made in rock crystal in more abstract and more restrained shapes were also in fashion. Cases in the shape of elongated octagons, quadrilobes and stars or circular and cut into lobes were the most current.

Diamonds, rubies, sapphires and emeralds often went into the decorative composition of luxury watches. The jewelry inventory of Elizabeth I, done in 1587, mentions several watches ornamented with precious stones.[18] Gabrielle d'Estrées, one of the mistresses of Henri IV of France, owned a watch similar to those of that queen. Her inventory, drawn in 1599, describes it thus: "very beautiful watch, with a quantity of diamonds and, hanging down, a pear-shaped pearl, worth 700 escus."[19]

87

◁ 87 Tulip bud-shaped watch. Enameled gold case. Movement signed "J Jolly A Paris." First half of the seventeenth century. L. 30; W. 26. This piece is an example of the very refined use of champlevé enamels in watch decoration. Its movement can be attributed to Josias Jolly, who was received as a master watchmaker in 1609 and died in 1642.
Paris, Louvre, Garnier Coll. inv. OA 7032

88 Fleur-de-lys watch with its red leather protective case and its crank key. Polished silver case. Movement signed "J Dracques A Nérac." France, first half of the seventeenth century. D. 37.
Paris, Louvre, Garnier Coll., inv. OA 7047

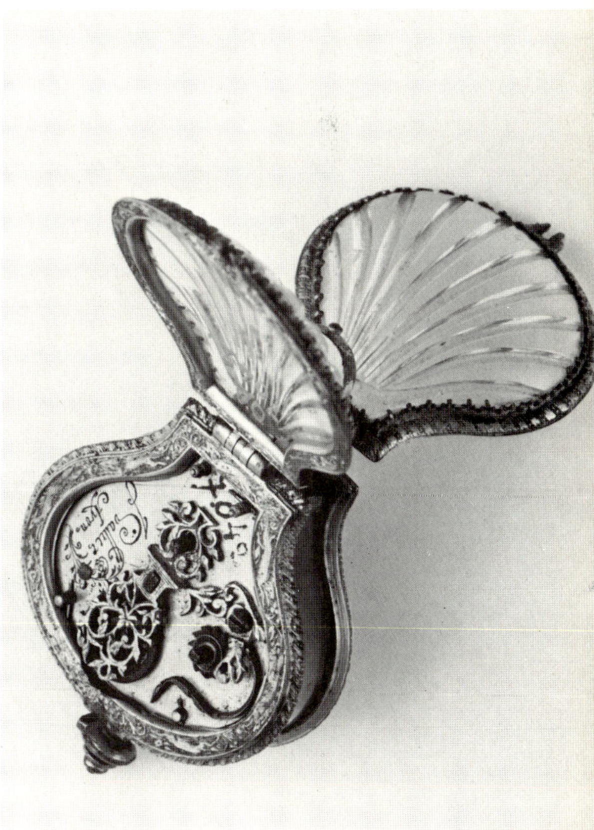

89 a,b. Shell-shaped watch. Rock-crystal case. Movement signed "Jean Vallier A Lyon." First half of the seventeenth century. L. 35; W. 35. Standing on a gilt brass plate engraved with foliage, the dial is decorated with a goat and trees in translucent enamel. The movement has a cock and pawl engraved and pierced with foliage.
Paris, Louvre, Garnier Coll., inv. OA 7042

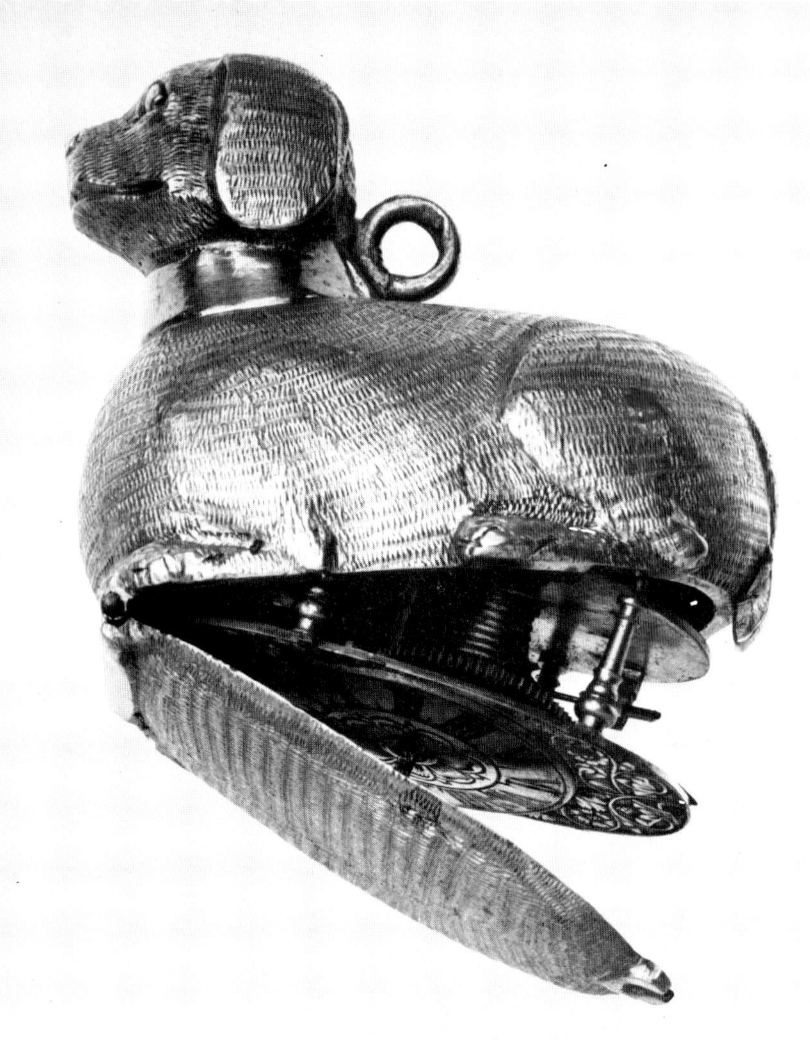

90 Watch shaped like a dog. Movement signed by Jacques Joly. Geneva, mid-seventeenth century. L. 41.5.
Jacques Joly (1622–1694) became a master around 1645.
London, British Museum, inv. 88, 12–1, 206

91 Rabbit-shaped watch. Movement signed by Pierre Duhamel. Geneva, third quarter of the seventeenth century.
Born around 1630, P. Duhamel lived in Blois before taking refuge in Geneva in 1654. He died in 1686.
Geneva, Musée de l'Horlogerie et de l'Emaillerie, inv. AD 198

92 Poppy bud watch. Amber case. Anonymous movement. End of the seventeenth century to the early eighteenth. Th. 26; W. 31.
Through the flower the round silver dial is visible, protected by rock crystal held by six fleur-de-lys claws.
Paris, Louvre, Garnier Coll. OA 7071

Again, let us quote an interesting piece of evidence provided by a legal act from the French National Archives. On November 19, 1569, Yves de Brynon, lord of Sardes and Guyencourt, gave for payment of part of his debts "a watch-clock set in crystal, garnished in gold through the garnitures, with diamonds and rubies all around and a pearl hanging down. . . ."[20]

More rarely, hard stones or hollowed stones housed watch movements. An egg-shaped agate watch is indicated in the jewel inventory of Queen Elizabeth. Very few of these jewels have managed to reach us. A watch signed Jean Barberet of Paris, is an example of their style: the movement is placed inside a rounded emerald; the open-work lid carries an emerald through which the hand can be seen (private collection). To this kind of watch made in hard or precious stones, one can add a watch whose amber case is hollowed out and cut in the shape of a six-petaled poppy, kept in the Paul Garnier Collection (inv. OA 7071).

Puritan watches (1640–1660)

In contrast to the fantasy and decorative richness of the form watches, a type of watches today called the Puritan watch appeared in England, around 1640, surviving through the Cromwellian interregnum (1649–1660). While an answer to the principles of temperance and austerity of the Puritans of the time, it also reflected the severity of the rules imposed by Cromwell in reaction to the Stuarts' taste for luxury.

The Puritan watches are characterized by polished, undecorated metal cases. Their dials are seldom decorated; the hour circle is visible under a circular crystal glass maintained by an inside mounting fixed by two screws. The pendant, a round ball, gives free passage to the suspending ring. Two watches very representative of the style are in the British Museum. Both bear the signature

93 "Puritan" watch. Polished brass case. Movement signed by Robert Grinkin. London, mid-seventeenth century. L. 36; W. 30.
London, British Museum, inv. 1786, 9–26, 1

of Robert Grinkin, a master in the Clockmakers' Company of London in 1648 and then in 1654.[21] Oval, made of gilt brass, without any ornament, these two watches have a protective brass case (inv. 1786, 9–26, 1 and 74, 7–18, 47).

Round watches in engraved metal (1580–1660)

As a prelude to the triumph of the round watch, the first flat and circle-shaped watches present a bulging band evoking a tire. Their engraved decoration is similar to that of the drum watches. Like those, they have an openwork lid. When the movement is equipped with a striking system, the bottom and the periphery are pierced. A molded stud is sometimes affixed opposite the pendant.[22]

In the second quarter of the seventeenth century, a form of watch close to the future pocket watch succeeded the circular watches. The band became rounded to prolong the outward curves of the lid, the "dome" and the flattened bottom. The openwork lid was replaced by a full lid or a glass framed in a bezel.

This kind of case, often made in silver, is usually ornamented with a stylized or naturalistic floral decoration. The bottom may be finely engraved and chased with an interlacing of flowery foliage where strawberry flowers, tulips, daisies or roses predominate in the style of Gilles Légaré's models. Etienne Hubert's watch in the Garnier Collection (inv. OA 7046), the one of Jacques Hubert of the Musée International d'Horlogerie (inv. 560) and the one of Edward East, which, according to tradition, belonged to Charles I and is now in the Victoria and Albert Museum (inv. 64–1952), are thus decorated with much refinement.

94 Round watch. Engraved and pierced silver case. Movement signed "Eduardus East Londoni." London, mid-seventeenth century.
Made by Edward East, watchmaker to Charles I, the movement has an alarm whose dial, numbered in Arabic numerals, is decorated with flowers in a naturalistic style.
London, Victoria and Albert Museum, inv. M 64–1952

95 Octagonal watch. Rock-crystal case. Movement signed by Michael Nouwen. London, around 1600. L. 35; W. 30.
The brass dial, decorated with champlevé enamels, can be seen through the rock-crystal lid.
London, British Museum, inv. 88, 12–1, 189

96 *Left.* Oval watch. Enameled gold case. Second quarter of the seventeenth century. The movement by A. Dobson is at the back (London, eighteenth century).
This piece is a good example of the very refined use of enameling over gold in high relief. It was probably offered by Charles I to his counselor, William Graham, count of Menteith (1591–1661).
Center. Round watch. Enameled gold case. Movement signed "Eduardus East Londini." London, mid-seventeenth century.
Very elegant, this case is decorated on the outside with embossed white florets, standing out from an opaque blue enamel base. Inside, it is ornamented—as is also the dial—with paintings on enamel.
Right. Round watch. Enameled gold case. Mid-seventeenth century. The movement is at the back.
The case is decorated with champlevé enamels composing a naturalistic floral ornamentation, such as can be seen on the engraved cases.
London, Victoria and Albert Museum, inv. 446–1884, 14–1888, M81–1913)

94

95

96

Other cases offer a large flower with open, flat petals, surrounded either with flowery foliage or with radiating gadroons. The first composition is found on two German watches of the Olivier Collection (inv. OA 8300 and OA 8301),[23] whose cases are pierced completely through to let one hear the striking movement. The second composition was adopted for a case containing, for instance, a movement made by Grégoire, a Blesian watchmaker[24]; the gadroons on that watch are engraved with fleurons, but they could also have been smooth.

The harmonious shape of these round watches is underlined by the decoration of the band, the dome or the bezel, which extends that of the bottom. Moreover, the decoration of the band often repeats on a lesser scale the decorative composition of the case.

Obvious analogies between the engraved decoration of these cases and the enameled composition of the cases produced in the same period are noticeable. Like the engravers, the enamelers used the floral and foliage decoration models proposed by ornamenters like Jacques Hurtu and Michel Le Blon in the first quarter of the seventeenth century, then by Jacques Vauquer and Johann Paul Hauer around the middle of the century.[25]

Watches ornamented with enamel

During the sixteenth century and until the start of Louis XIII's reign (r. 1617–1643), enamel played a leading role in jewelry composition. Even in the most splendid pieces, the stones were enchased, isolated in large enamel montures. It is then unsurprising that the taste for enamel was also seen in watch decoration, the watch case following the current fashions in jewelry and goldsmithing work.

The varied techniques used in the enameling art in the sixteenth and seventeenth centuries were applied to cases. A large number of extant pieces have a decoration uniting several techniques.

Champlevé enamels

Preferred to cloisonnés because of their easier execution, champlevé enamels were first used in arabesques and scrollwork compositions, later in more naturalistic floral decorations. The enameler would dig out with his graving-tool a gold or brass plate, taking care to preserve the partitions intended for edges in the decorative motives. Then, after depositing the translucent or opaque enamels in the compartments thus obtained, he would fire the piece.

The dial was the first part of the watch to be ornamented with champlevé enamel. Sixteenth-century German watches often have a dial decorated in the center with a sun in red and black enamels of this kind. At the end of the century, the dial of rock-crystal watches was sometimes enhanced with enamel, representing flowers, arabesques or small animals.

Some of these rock-crystal watches show, besides a champlevé enamel dial, a plate also enameled. Visible under the rock crystal, these enamels, with their bright colors, accentuated the refinement of the case. An hexagonal watch with an unsigned movement, in the Louvre, presents a dial and a plate thus strewn with enameled flowers, dragonflies and garlands (collection Garnier, inv. OA 7059). A watch in the British Museum (inv. 88, 12–1, 189), its case cut from a mass rock crystal and without armatures, has a gilt brass dial ornamented with champlevé enamels forming a sun in the center and, on the periphery, foliage and flowers. The plate of the movement, signed "Michaell Nouwe" is protected by another plate decorated like the dial. In the Cluny Museum, there is a watch signed by Duboule that is comparable to these two pieces, in the shape of a star, richly studded with diamonds, with both the dial and the plate covering the plate of the movement, decorated with scrollwork and red and green enamels on a ground of black enamel.[26]

As in jewelry of the same period, enamels are associated with gems in watch decoration. A magnificent example of that refined association is the oval watch in the British Museum, made around 1620, the bottom of which is formed of a large cabochon sapphire, while the lid is decorated with another cabochon sapphire, surrounded with eight smaller sapphires. The inside of the lid, the dial and the armatures are enameled with an exquisite delicacy, with refined foliage, birds with chatoyant plumage and insects with glistening wings, in a naturalistic style over a blue and white ground.[27]

Despite the infatuation for painting on enamel in the mid-seventeenth century, champlevé enamel cases were still created during the period. Their floral decoration relates them to the cases engraved or painted "pell-mell" after the models of Jacques Vauquer

97 Oval watch. Engraved and enameled case. Movement signed "Aug Buschman." Augsburg, end of sixteenth century. L. 84; W. 51; Th. 40. The lid of the watch is ornamented with a Limoges painted enamel, remarkable for the richness and the depth of its colors. The monogram *IL* painted between Christ's feet permits its attribution to Jean Limosin. La Chaux-de-Fonds, Musée International d'Horlogerie, inv. 1147

and Gilles Légaré. A pretty watch case of that kind, containing a posterior movement, is kept in the Victoria and Albert Museum (inv. M 81–1913): a reserved gold *résille* emphasizes the asymmetrical decoration of the flowers and leaves on the cuvette.

Limoges painted enamels

A few preserved pieces give evidence of the application of enamel painting in watchcase decoration. Some of them prove, through the signatures they bear, that their cases were ordered by foreign watchmakers from enamelers in Limoges.

The Musée International d'Horlogerie has in its collection an oval watch (inv. 1147) whose case has a lid painted in Limoges enamel signed "J. L." (Jean Limosin), representing Jesus carrying his cross. The silver bottom of the case is engraved after a work by E. Delaulne. The movement is signed "Aug. Buschman." The abbreviation "Aug." stands for the city of Augsburg, where an important watchmaking family named Buschman worked during the sixteenth and seventeenth centuries. The signature of the famed Limousin enameler suffices to prove the French origin of the case. Such pieces give weight to the hypothesis that numerous watchcases were engraved or enameled in France, and then exported.[28]

In the Pierpont Morgan Collection,[29] an oval watch made around 1610–1625 is also decorated with Limoges enamels. The scenes are inspired from Ovid's *Metamorphoses*; one of them, signed "I. R.," permits the attribution of these compositions to Jean or Joseph Reymond. The movement bears the famous signature of David Ramsay. This watchmaker of James I seems often to have called on the service of French artists; we have already mentioned two movements that he created, contained in cases signed by the Blesian engraver Gérard de Heck.

Watches are preserved that reveal enamels in relief, a technique also developed notably in Limoges, which were mainly used around the mid-seventeenth century in flower decorations. A round watch with the movement signed "Eduardus East Londoni," kept in the Victoria and Albert Museum, has an exquisite case ornamented with white florets in relief standing out on a sky-blue background (inv. 14–1888). The British Museum keeps a circular watch, also dating from around 1650, decorated with diamonds and slightly relieved enamel flowers on a black enamel background; its movement is signed by an immigrant French watchmaker, D. Bouquet (inv. 88, 12–1, 219). The group of cases decorated with a medallion painted over enamel, usually presenting the bust of a plume-helmeted woman, made around the mid-seventeenth century, is often framed with enameled flowers or foliage in relief.

98 *Left.* Six-lobed watch. Enameled gold case. Movement signed by L. Vautyer. Blois, early seventeenth century. D. 38.
The case, pierced in the shape of a rose window, is decorated with polychrome enamels.
Right. Round watch. Enameled gold case. Movement signed: "D. Bouquet Londini." London, mid-seventeenth century. D. 46.5.
The case is ornamented on the outside with a naturalistic embossed enameled floral decoration over a black enamel base. On the lid, diamonds enhance the decorative richness.
London, British Museum, inv. 92, 5–23, 1 and 88, 12–1, 219

Enameling over gold high relief

The enameling of pierced metal cases forms another type of decoration used in the first half of the seventeenth century. The foliage and the flowers cut out of the metal were covered either entirely or partly with translucent or opaque enamels. An oval watch of the Victoria and Albert Museum (inv. 446–1884) is decorated using this process. Its gold case is pierced with flowery foliage, enhanced with transparent red and green, opaque blue and white enamels. The Blesian watchmaker Louis Vautyer signed a watch in the British Museum (inv. 92, 5–23, 1) ornamented with delicately cut rosace, partly underlined with red and green enamels.

Around the mid-seventeenth century, enamelers applied the technique to a more naturalistic flower decoration that is also found on engraved metal and enameled, unpierced cases. The Victoria and Albert Museum has a round case of this kind, inspired from Jacques Vauquer's models, pierced with large enameled flowers (inv. 2362–1885). The watch by Edmé Burnot in the Musée International d'Horlogerie, decorated with the portraits, after Velázquez, of Felipe IV and Anna-Maria of Austria, presents similar ornamentation. Both medallions, painted on the bottom and the lid, are framed with a garland of flowers in full bloom, pierced and enameled in bright colors.[30] This kind of enameling on metal cut in the shapes of flowers or knotted ribbons is also found in the composition of necklaces, bracelets and pendants of the same period, as well as in the mountings of hard stone vases.

99 *Left.* Gold enameled round case. Mid-seventeenth century.
This piece offers an example of enameling over high-relief gold, applied to a naturalistic floral decoration, similar to the one shown in the watch decorated with the portrait of Philip IV of Spain (ill. 101).
Right. Octagonal watch. Enameled gold case. Movement marked "A Bloys". Early seventeenth century.
The case is delicately ornamented according to the so-called enamel *en résille* on glass technique.
London, Victoria and Albert Museum, inv. 2362–1855, 2553–1855

98

99

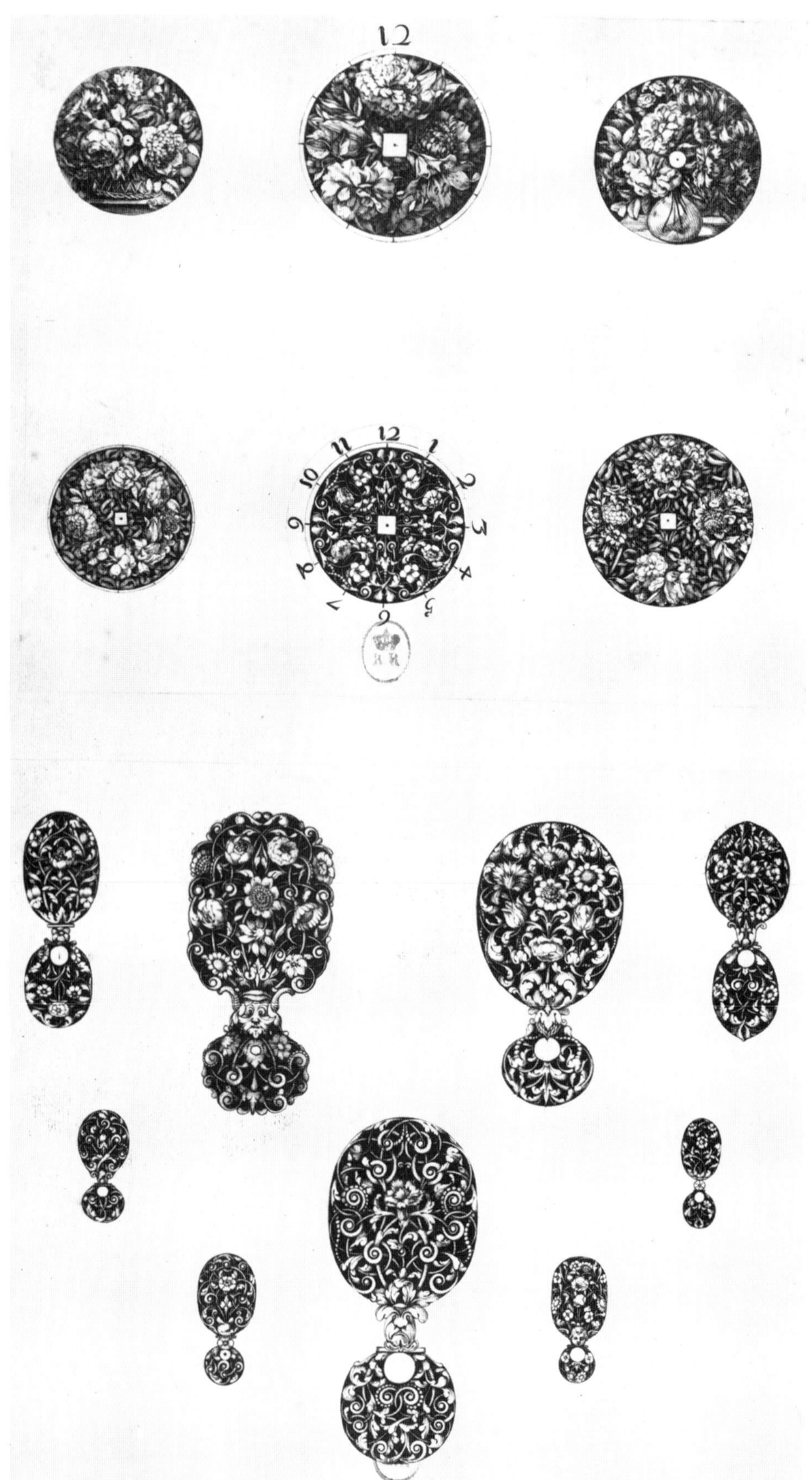

100 Case and watch-cock models engraved by Jacques Vauquer. France, second half of the seventeenth century.
Numerous engraved and enameled cases still extant reflect the naturalistic style of these models.
Paris, Bibliothèque Nationale, Cabinet des Estampes (Le 49 rés., p 10)

101 Enameled gold round watch. Movement signed "Edme Burnot A Bruxelle." Spanish Low Countries, around 1665. D. 60; Th. 25.
The portrait of Philip IV (1605–1665) wearing the Golden Fleece decorates the lid. This effigy of the Spanish sovereign is identical to the one engraved in 1663 by Baltazart Moncornet from a probable painting by Velazquez, now lost. The medallion is surrounded by a garland of enameled flowers in the style of Jean Vauquer.
La Chaux-de-Fonds, Musée International d'Horlogerie, inv. 1120

102 Models of oval cases, engraved by Stephanus Carteron. France, 1613.
This decorative style can be found on cases ornamented with champlevé enamels or *en résille* over glass enamels.
Paris, Bibliothèque Nationale, Cabinet des Estampes (Le 48 rés., p 21)

"En résille enamel on glass"

A very delicate technique, known under the name of "*en résille* (hairnet) enamel over glass," was applied to watch decoration. The rare cases preserved giving evidence of that technique also provide an eloquent testimony of the skill of the goldsmith-enamelers of the seventeenth century.

An octagonal watch decorated according to this process, whose movement bears the inscription "A Bloys," is kept in the Victoria and Albert Muscum (inv. 2553–1885). Gold foliage work, punctuated with red fleurons and green leaves, creates a symmetrical composition forming a network in dark blue glass. The enamel was applied inside the scooped out metal. An oval case, also made by a French enameler, around 1630, now in the British Museum, presents a similar decoration. The Louvre is also one of the very few museums owning a watch made using that complicated technique (inv. OA 683).

Translucent enamels

As has been stated many times already, the combination of opaque and translucent enamels was done routinely. Translucent enamels accented the precious refinement of the cases, their luster imitating that of colored gems. A very successful example is the tulip-bud watch kept in the Soltykoff Collection, now in the Garnier Collection (inv. OA 7032). The three petals forming the case are decorated, on a white enamel ground, by reserved gold foliage, enhanced by translucent red enamel flowers and translucent green enamel leaves outlined by thin reserved gold partitions.

Translucent enamels were also stretched over larger surfaces at times to be used as a ground for chased gold ornaments. In this case, the sheen of the enamels was set off by a hand-made guilloche on the metal they covered. The square watch whose movement was signed by the Blesian watchmaker Nicolas Lemaindre, in the Paul Garnier Collection, is enameled this way on all its faces with translucent green on a guilloched ground, throwing into relief reserved and engraved gold foliage.[31]

Having displayed their talents in the traditional enameling techniques, French enamelers successfully exploited a process that revolutionized the decoration of watches: the painting on enamel.

Watches ornamented by painting on enamel (1630–1670)

Appearing in the second quarter of the seventeenth century, painting on enamel knew a rapid success that lasted for more than two centuries. This success was largely due to the fact that, because of their lasting toughness and inalterability, the pieces achieved through the new technique advantageously replaced miniature painting, then very much in fashion in the decoration of jewels or small objects for personal use, which suffered from exposure to light and the elements.[32] Painting over enamel found an immediate field of application in watch decoration.

The beginnings of painting on enamel are closely linked to watchmaking history. Goldsmith-enamelers and Blesian watchmakers united their talents and produced a new type of round, fairly flat watch, of which the large and almost level surfaces of the lid and the bottom were entirely painted. This kind of watch, known as the "pan" or "basin" watch, increased the repute of the Blesian craftsmen and contributed through its vogue to the development of the horological industry.

Origins of painting on enamel

The origins of the new technique are obscured because of extreme scarcity of contemporary documents that mention it and the anonymity of most of the paintings on enamel.

Until the present day, the only reference that historians of enamel work were able to use was the testimony, even though a late one, of Félibien, in his *Principes de l'architecture, de la sculpture, de la peinture et de tous les autres arts qui en dépendent* (Principles of architecture, sculpture, painting and all the other arts dependent on them), published in 1676: "Before 1630, these kinds of work were still unknown, for it was only two years after that Jean Toutin, goldsmith in Châteaudun, who enameled perfectly well with ordinary enamels and the transparents and who had a certain Gribelin as his disciple, having started to search for the means of using enamels that might produce mat colors to make various hues, which might completely melt when fired and keep the same evenness and the same gloss, finally found the secret that he communicated to other workers, all of whom then contributed to rendering it more and more perfect" (p. 423).

In the same year, the *Journal des Sçavans* confirmed Félibien's testimony: "A Frenchman named Toutin, a watchmaker from Châteaudun . . . found, about the year 1632, the secret of thick and opaque enamels with which portraits can be made as well as in oil. . . ." (September 14, 1676, p. 118). Around the same period, scientist Philippe de La Hire was writing a small report entitled "De la Peinture en émail" (On painting on enamel), published in Volume IX of *The History of the Royal Academy of Science*, 1666–1699 (pp. 728–730), in which he noted, following Felibien's assertions: "Enamel painting is an entirely modern invention owed to France, and principally the high degree of perfection to which it has been brought in the past 50 or 60 years. . . ."

If the documents found in the Blesian archives by Abbé Develle are considered, the date of 1632 for this invention given by Félibien and accepted by contemporary writers seems rather late. Various legal acts indeed reveal that as early as 1630 the technique was already widespread in the goldsmith-enamelers' shops. H. Clouzot aptly emphasizes: "1632, given by Félibien probably according to informations furnished by Toutin's descendants, can be considered only as an extreme date, indicating the start of an active exploitation of the process. There had been certainly a dozen years, maybe more, beforehand of groping research and experiments."[33]

Jean Toutin, a son of goldsmith Etienne Toutin and Marie Vallée, was born in Châteaudun in 1578 and received his baptism in the Reformed religion. He was established in Blois as a goldsmith in 1604. After his marriage, which took place on June 14, 1609, he returned to his native town to open his shop.

We have found a document showing the presence of Jean Toutin in Paris in 1623. A judgment from the bailiff court, dated September 16, 1623, forbids "the said Jean Toutin, Master Goldsmith, ever to undertake any other enterprise in the watchmaking trade. . . ." and decrees the confiscation of watches and clocks seized in his shop.[34] This act is evidence of his stay in Paris before 1630 and 1632, dates in which the registry of the Châteaudun Reformed church carries the death certificates of two of his children. It also proves, moreover, the master's tight links with the profession of horology.

Jean Toutin died on June 14, 1644, leaving a daughter and three sons, two of whom, Henri and John, were enamel painters.

Without Félibien's historical notation, we would have ignored his innovative role, which no contemporary document mentions and no signed work reveals. We know only of his two series of printed engraved plates, made in 1618 and 1619 in Châteaudun, proposed as goldsmith's models and decorated with delicate foliage in the traditional style of such contemporary masters ornamenters as Jacques Hurtu.

It is difficult to define the share of Jean Toutin and other goldsmiths in Blois in the invention and elaboration of the technique of painting on enamel. According to Schneeberger, "painting over enamel is not . . . properly speaking, a discovery of goldsmiths from either Châteaudun or Blois, but the application by these goldsmiths of a process that, for at least a century, had already been known by Limousin enamelers."[35] His demonstration is founded in particular on the technique employed in the portraits of Léonard Limosin; in order to bring out every nuance of the modeling of the hands and face, the famous enameler was already using stipple work on a white enamel ground. H. Clouzot also looked for precedents to the technique of enamel painters: he discovered them notably in plaquettes covered with an opaque enamel and decorated with outlined figures.[36]

At any rate, it would be an injustice to minimize excessively the originality of Jean Toutin's experiments, for no one before him had succeeded in producing a range of vitrifiable colors comparable to that of oil paints, whose application with a brush would permit the notation of the smallest details. Félibien rightly points out the two important innovations of the first enamel painter: in the sixteenth century "all the enamel works, as much over gold as over silver or brass, were ordinarily only clear and transparent Enamels. And when thick Enamels were used, each color was laid flat and separated. . . . But the manner had not been found to paint as is done today with thick and opaque Enamels, nor the secret to compose with them the Colors we use now" (*op. cit.*, p. 420).

The novelty of the technique is also stressed in the oldest treatise we have, which exposes its secrets. Published recently,[37] this manuscript, entitled *Des Emaulx* (Enamels), in the Sloane foundation (1990) of the British Museum, is quite important since it was unquestionably written by Petitot and Bordier, between 1638 and 1644. The new way of painting is described there as a "very rare and curious method to use Enamels as well as Color in Illumination."

Technique of painting on enamel

The first treatises revealing the secrets of painters over enamel were published in the eighteenth century; in the preface of his treatise appearing in 1721, under the title "L'art du feu ou de peindre en émail" (The art of firing or painting in enamel), Jacques-Philippe Ferrand[38] regrets that painters who had excelled in this art did not write anything about it, "Either because they did not feel sufficient enough in the art of writing to express themselves on it, or they had in their family a Successor to their science, to whom they did not wish to do wrong, or else that they had only the colors, which they did not know how to make, except for a few, as I heard from themselves & from the S[ieu]r Trocus, a learned chemist of their time." The meticulous description given by Durand, the "painter to the duke of Orléans," in the article "Emaillerie" (Enamel work), in the *Encyclopédie*[39] and in a *Traité des couleurs pour la peinture en Email et sur la procelaine. Précédé de l'art de peindre sur émail* a posthumous work by d'Arclais de Montamy (1765), provides ample information on the process of painting on enamel and stresses the difficulties met in the course of their work by the enamelers.

The gold or brass plate[40] upon which the enameler intends to paint must be lined with a fillet to hold the enamel or be raised by champlevé to the dimensions of the painting when the enamel does not cover it completely. To let the enamel take hold, the painter distributes scratches and hatchings on the plate, then ready to receive the opaque, generally white, enamel of the ground, more or less soft according to the quantity of flux it contains. The piece undergoes its first firing. "They have long, flat tongs they call *releve-moustaches* (mustache-raisers), which they use to lift up the plate & bring it to the fire," Montamy explains. After being withdrawn from the fire, the plate receives a second load of enamel and a second firing; two loads are enough for gold; three are recommended for brass. The piece must be counterenameled according to the same process; any piece enameled "in full" on the side to be painted is counterenameled on the other side, half as much so if it is convex. The operation particularly concerns the watchcases of the seventeenth century and the first half of the eighteenth, the bottom of which is enameled in full; later when the enamel will occupy only a limited space on the bottom of the case, there will be no counterenamel.

After the preparation of the base meant to receive the paint and the counterenameling, the enameler prepares his colors, vegetable or mineral oxides that he reduces to a very fine powder. To make sure of the quality of his colors, the painter took care to prepare small enameled plates he had numbered and which are called "inventories." This way, he forms his palette and controls the effects of the fire on his colors. After this preparation, he draws his piece with Mars rouge, then slightly colors it and harmonizes the values in a light tonality with colors he mixes with a flux such as lavender oil, which helps as a vehicle and a revealer. The plate is fired. The painter, above all, cannot miss the moment when it must be withdrawn, at the point of melting. According to Montamy: "This maneuver is very critical; it keeps the Artist in great anxiety; he is not ignorant of the state in which his piece was put on the fire, nor of the time he spent in painting it; but he does not know at all its condition when he draws it out, & whether he will have wasted in one moment the assiduous work of several weeks. It is in the firing, it is in the furnace that are manifested the bad qualities of the coal, metal, colors & enamel, pitted surface, swellings, even cracks. A sudden flash of the fire will sometimes erase half the paint, & of a whole tableau well worked on, well harmonized and well finished, there remain on the base only feet, hands, heads, scattered and separated limbs, the rest of the work having vanished; that is why I have heard Artists say that the time of firing, however short it may have been, was almost a time of fever that wearied them more & was more injurious to their health than whole days of continuous labor." Several firings are necessary to finish off the enamel; the painter is careful to superimpose layers of increasingly accentuated colors and to save the delicate colors for the last fire.

103 Medallion painted on enamel, signed and dated "Henry Toutin, M^e^ orfèvre à Paris, fecit 1636." L. 45; W. 32.
London, British Museum, inv. BL 3470

104a,b. Round watch. Painted enamel case by Henry Toutin. Movement signed by Antoine Mazurier. Paris, around 1641.
The light colors and stipple technique characterize the historical and allegorical paintings of this exceptional piece, which was made about ten years after the invention of painting on enamel by the artist's father.
Amsterdam, Rijksmuseum, inv. NM 638

103△

104a

104b

Enamelers of Blois

The activity of Blesian enamelers in the art of painting over enamel is mostly known through archival texts discovered by Abbé Develle. As noted before, some of these documents show that the date 1632 given by Félibien was late. The new technique had been routinely used by then, as proved by a judicial paper of July 29, 1632, discussing an order given by the watchmaker Lemaindre to the enameler Christophe Morlière. The "painted enamel" case delivered by Morlière was refused by Lemaindre; Isaac Gribelin, a "disciple"—according to Félibien—of Jean Toutin, was called in as an expert and declared that the work could not be considered a painted enamel case as the contract specified and as Morlière was in habit of making.[41]

The speed with which the process had been introduced in the workshops of Blois is the more surprising if one considers the mystery with which the old masters surrounded their discoveries. According to Félibien, Jean Toutin himself had communicated his secret "to other workers." Indeed, it seems that Jean Toutin was not very jealous of his inventions, since he was assisted in his experiments not by an apprentice, but by a "disciple," Isaac Gribelin, himself a goldsmith, already well established in Blois, where he had received his mastership in 1618.

Isaac Gribelin, a protestant like the Toutins, belonged to a Blesian family, several members of whom acquired reknown in watchmaking, jewelry and ornamental engraving. He himself became famous thanks to his talent as an enamel painter. In his *Histoire de Blois* (History of Blois), published in 1682, Bernier praises "his genius so particular for portraits that he was one of the first men in his time in that practice, succeeding equally in enamel and pastel" (p. 73). In 1634 his reputation won him his position as a juror in the Goldsmiths' Guild. Despite the interdictions stated in the guild regulations with the object of preserving the rights of the watchmakers, his business sense led him to organize the production of watches whose movements were secretly made in his shop. This illicit dealing was uncovered in June 1636; he was ordered to dispose of all the watches thus made, selling them outside the city, and forbidden to repeat the offense.[42] He died in 1651, leaving, it seems, only daughters.[43]

Christophe Morlière certainly was Isaac Gribelin's most talented rival. Born in Orléans on May 12, 1604, a Catholic, he settled in Blois, where he married on February 11, 1630, Marie Poëte. At the time of his wedding, he already held the envied title of goldsmith and engraver to Gaston d'Orléans, the brother of the king. Already at the time, he had joined his talent as an engraver to that of a painter, as revealed by his disputes in 1632 with Lemaindre over a case decorated with flowers. His well-reputed shop occupied numerous apprentices and journeymen, who filled it "up to the attics."[44] Morlière's reputation continued to increase until his death in 1644. Thus, having decided to offer a splendid enameled watch to the duchess d'Orléans, Marguerite of Lorraine, on the occasion of her return to France in 1634, the Corps of the city of Blois turned to him: "In accordance with what had been formerly discussed to make a present to My Lady the duchess d'Orléans, it is resolved that the city will have made by Morlière a gold watch with a case enameled with figures and personages and to use for it up to the amount of XII^{c} 1. or any other such sum as Messieurs the Aldermen will consider."[45] No work has survived bearing Morlière's signature. The painted-enamel floral decoration of a watch now in the Imperial Treasury in Vienna is attributed to him, for the movement has the signature of his father-in-law, Jacques Poëte.

A pupil of Morlière's, Robert Vauquer proved himself so skillful in the art of painting on enamel that his reputation equaled that of his master. Born in Blois on April 31, 1625, he was a brother of an illustrious engraver, Jacques Vauquer (1621–1686). He died on October 20, 1670. Collections preserve a few paintings signed by him, featuring portraits or religious scenes. The Vatican's Christian Museum has an eighteen-scene set of the Life of Christ, signed "opus Roberti Vauquerii Blesensis an 1660."[46] Two signed portraits of women, one dated 1664, the other 1670, are pointed out by Henri Clouzot[47] in the collection of the count of Ilchester. In the collection of the duke of Rutland, the "Slaughter of the Innocents," after Raphael, is signed "Vauquer Pin." Paintings signed "Vauquer Fec," inspired by Jules Romain's "The Battle of Constantine," decorate a watch in the Piane collection.[48] Clouzot attributes to Vauquer a watch in the Paul Garnier collection (inv. OA 7076), decorated with an "Adoration of the Magi" that reproduces a work of his brother, Jacques.[49]

Pierre Chartier belonged to the same generation of Robert Vauquer. Born in Blois on January 5, 1618, he began his career as a master goldsmith. In 1651 he was already living in Paris, where he acquired a reputation as an enamel painter, as Félibien testifies in his *Principes* (p. 424): "Pierre Chartier from Blois started to make flowers, in which he succeeded perfectly, and immediately several people were seen attaching themselves to that way of painting, with which quantities of medals and other small works were made." No work can, with any certitude, be attributed to him.

Many other enamelers whose names have been lost contributed between 1630 and 1640 to perfecting the art of painting on enamel. One can, however, cite the name of Dubie who, according to Félibien,[50] worked as a goldsmith in the Galeries of the Louvre and was one of the first to know the technique of overenamel painting.

Jean Toutin's sons

Born on July 28, 1634, Henry Toutin started a brilliant international career as an enamel painter from the time of his reception to his mastership, in 1636 or 1637, as a goldsmith. Established in Paris with his family, he executed two works in 1636 in a rather awkward technique; today, they are the two oldest pieces painted on enamel that we know. One is an oval medallion, representing "Diana and Acteon" and a sea battle ("The Taking of Constantinople"), signed: "Henry Toutin, M^{e} orfèvre à Paris, fecit 1636" (British Museum, inv. BL 3470). The other one is a portrait of Charles I, kept in the Rijksmuseum, signed "Henry Toutin, orfèvre à Paris a fait cecy l'an 1636."

His technique was perfected in a few years, as proven by the splendid watch made on the occasion of the wedding of William of Orange and Mary Stuart in May 1641, twice signed and now in the Rijksmuseum in Amsterdam. The paintings, in which history mingles with mythology to serve the glory of the princely couple, are in the style of baroque allegory, made fashionable in Northern Europe by Rubens. On top of the lid, the prince discovers his future wife driving a chariot pulled by two leopards; inside, the wedding ceremony unfolds in a temple. In the back of the cuvette, Neptune, brandishing his trident, drives the chariot bringing the newlyweds toward the United Provinces, personified by five females; inside, appearing like Mars and Venus, the prince, armed with his cuirass, and his spouse, draped in an antique style, gaze on a battlefield on which a temple to Victory is erected. The band is enameled with a view of London and one of the Hague. On the dial, cupids offer flowers to the couple. The movement is signed "Antoine Mazurier à Paris." In the following years, royal commissions kept on coming. According to Félibien, "after the death of the late King Louis XIII, he made for the queen regent a golden watchcase, enameled with white figures on a black background."[51]

Henry Toutin's talent was varied; he also knew how to paint floral decorations, much in fashion in the first half of the seventeenth century, with exquisite delicacy. An enameled medallion in the Kunsthistorisches Museum in Vienna, signed "H. Toutin," is sensitively ornamented with tulips, peonies and carnations, among which smaller flowers, pansies and daffodils, stand out. It contains the portraits of Louis XIV as a child and Anne of Austria, painted over vellum and thus can be dated between 1642 and 1643.[52] We still have among his work, a *Portrait of a Woman* signed "Henry Toutin fecit 1651" (Victoria and Albert Museum), *Venus and Mars*, signed and dated 1662 (Bayerisches Nationalmuseum), and the famous *Tent of Darius* made in 1672 from Lebrun and kept at the Musée de l'Horlogerie, in Geneva. This masterpiece, dazzling in the subtlety of the colors and the quality of the drawing, was perhaps ordered by Louis XIV. His death or exile for his religious convictions, explains the absence of his name on the lists of Parisian goldsmiths established after 1683.[53]

His brother Jean, born in Châteaudun on November 14, 1619, also had a career as enamel painter. His signature is on two watchcases admirable for the brightness of the colors and the firmness of drawing. One, now in the British Museum, is decorated in the style of landscapes animated with bathing nymphs, which made the popularity of the Dutch painter Poelenburg in the mid-seventeenth century. The other one, belonging to the Louvre, is ornamented with scenes of bacchantes and satyrs (Garnier Collection, inv. OA 7075). Around 1645 he left for Stockholm, invited by Queen Christina. Some events unknown to us prevented his appointment as a painter on enamel to the queen of Sweden, given instead, in 1646, to Pierre Signac (1624–1684).[54]

The works: iconography and style

Few enamelers took the trouble to sign the paintings with which they covered watchcases. However, the decorations painted on enamel in Blois or Paris from 1630 to 1670 were executed by true artists who gave their art an unequaled perfection.

Two kinds of decoration dominated the production in that period: floral decoration and historical scenes.

The floral decoration, in fashion during the last third of the century, shows an obvious care for realism. The works in which it is represented–watchcases, but also pendants, medallions and book bindings–seem to have come from the color plates of an herbal. Executed with an extreme delicacy on a very light ground that makes the tiniest details stand out, the flowers are arranged in a seemingly free manner on the whole surface of the painting and clearly isolated from each other. The balance of masses and values is subtly respected; among tulips and carnations, smaller flowers such as daffodils and pansies stand out, sometimes with perched birds. The attention is wholly directed to the flowers themselves since the stems and leaves are barely sketched or are even eliminated.

Enamelers chose their models in print collections, such as J. Caillart's, *Livre de toutes sortes de feuilles pour servir à l'art d'orfébrie* [sic] (Book of all sorts of leaves for use in the art of goldsmithing), published in 1629, or F. Lefebvre, *Livre de feuilles et de fleurs pour servir à l'art d'orfèvrerie* (Book of leaves and flowers for use in the art of goldsmithing) (1635).

A beautiful example of that type of decoration is given by Henry Toutin's medallion, already cited, dated 1640. The watch of Jacques Poëte,[55] attributed to Morlière, also offers a floral decoration of the type. Two watches, in the Musée International d'Horlogerie, have a case in whose interior the cuvette and also the

105

105 Round painted enamel gold case by Jean Toutin. Paris about 1640. D. 47; Th. 20.
Mythological scenes ornament the exterior of the case, landscapes the interior.
Paris, Louvre, Garnier Coll., inv. OA 7075

106

106 Round watch. Painted enamel gold case. Movement signed "M Wentzel A Strasbourg." Mid-seventeenth century. D. 45; Th. 16.
The counterenamel of the bottom, the bezel and the center of the dial present a floral decoration, remarkable for the vivid luster of its colors and the realism in the drawing in the style of models engraved by Caillart and Lefebvre.
La Chaux-de-Fonds, Musée International d'Horlogerie, inv. 1331

▷

107 a–d. Round watch. Painted enamel gold case. Movement signed "F Baronneau A Paris." Mid-seventeenth century. D. 60; Th. 22.
Paintings on enamel inspired by the story of Venus decorate this piece, made around 1640–1650. Their style recalls French works of the first half of the seventeenth century, among which the enamelist certainly sought his models.
La Chaux-de-Fonds, Musée International d'Horlogerie, inv. 400

bezel holding the glass are sensitively and minutely ornamented with flowers painted on enamel. The floral decoration of both these cases is similar, as is also the style of the scenes on their bottom; they possibly come from the same shop, but it is risky to attribute them to Henry Toutin (inv. 1331 and 1219).[56]

About the mid-seventeenth century, another floral decoration, also in a naturalistic style, but marked with luxuriance, inspired by the models of Jacques Vauquer and Gilles Légaré, was the vogue. Abundantly "carpeting" the whole painted surface, flowers with petals opened wide crowd one another in apparent disorder. Roses, strawberry flowers, peonies and carnations dominate in these warm-colored compositions. A case painted in this style and containing a movement by John Adamson, is preserved in the Musée International d'Horlogerie (inv. 1116). This decoration was also used in garlands to frame medallions painted over enamel. An example of this is given by the watch in the Olivier collection that bears Denis Champion's signature (inv. OA 8333). The blue cameo medallion, showing Minerva on the bottom, is surrounded with a polychrome garland painted on enamel, composed of tulips, carnations and daffodils in the style of Vauquer.[57]

From the 1630s, enamelers largely exploited the possibility of executing real historical scenes in miniature, which the new process of painting with enamels was offering them. The "basin" watches, with their wide, flat surfaces, were indeed well conceived to receive paintings that were largely reduced versions of renowned works. Enamelers of the second third of the century received their inspiration in particular from Simon Vouet (1590–1649), the celebrated painter of Louis XIII. They also copied the works of Sébastien Bourdon (1616–1671) and those of artists among the French classicists in the 1650s, such as Laurent de La Hyre (1606–1656).

The Olivier Collection in the Louvre owns about one hundred of these watches decorated with paintings on enamel, the models of which have been for a large part identified. We invite the reader interested by the iconography of the decoration of enameled watches to refer there.

Whether mythological, romantic or allegorical, most subjects decorating the watches of that period represent famous lovers.

Among the couples of mythology, Venus and Adonis had a considerable success that, moreover, persisted for the whole last quarter of the seventeenth century with the paintings on enamel of the Huaud brothers. A case decorated around 1640 with episodes of their love is preserved in the Musée International d'Horlogerie (inv. 400). On the outside of the lid, Venus is at her toilet; a maid is cutting her hair, while Cupid presents her a mirror. Inside, the goddess is embraced by her lover. On the bottom, the three Graces and Love offer her flowers and a crown; inside, the tearful goddess discovers Adonis' dead body. The center of the dial shows Venus holding a flaming heart. Like this one, numerous cases made during the period, present several episodes of a same story unfolding on each of the faces of the case and on the dial.

The tragic loves of Paris and Helen, borrowed from the *Iliad* also inspired the enamelers. *The Judgment of Paris*, followed with *The Rape of Helen*, decorates many enameled watches around the mid-seventeenth century, such as the watch in the Olivier Collection, signed Charles Henou Embden (inv. OA 8317).

The tragic love of Dido and Aeneas was particularly favored. A painting by Sébastien Bourdon presenting *The Parting of Aeneas and Dido*, now in the Rouen Museum, was frequently copied in Blesian and Parisian workshops. It notably decorates two watches in the Olivier Collection, one signed by Elias Weckerherlin (inv. OA 8320), the other by Roger Dunster (inv. OA 8432),[58] and a watch now in the Rijksmuseum in Amsterdam signed "Riepott Ragensborg."

The love of Antony and Cleopatra has inspired the decoration of many cases. A Sébastien Bourdon painting, *The Landing of Cleopatra* served as a model for the enameling of a watch of the Olivier Collection (inv. OA 8331) and a Blesian watch of the Musée International d'Horlogerie (inv. 533).[59]

Paintings inspired by Heliodorus' novel *The Loves of Theagenes and Chariclea*, executed by Charles Poerson (1609–1667), decorate a relatively important number of watches in the second quarter of the seventeenth century still extant. One kept in the Olivier Collection, is signed by a Parisian watchmaker, Gamot (inv. OA 8318)[60]; another is signed "C. Bonnevye à Paris," in the Gélis Collection[61]; a third is described in the Emile Bloch Collection[62]; a fourth is in the Ermitage Museum; a fifth, with a movement signed "Willem Koster Amsterdam," belongs to the Patek Collection.[63]

The episodes taken from these famous love stories usually have as a subject an abduction, an embarkation or, again, a meeting at the time when one of the two protagonists disembarks from a ship. Thus, for a long time *The Embarkation of Chariclea* could pass for a *Rape of Helen*; the scene of *The Parting of Dido from Aeneas*, who makes ready to join his ship, has an iconographic form very close to that of *The Rape of Helen*. One thinks, for instance, of the one painted by Guido Reni. His *Meeting of Antony and Cleopatra* is not very far from the one of Aeneas and Dido.

Enamelers seem sometimes to have borrowed, either consciously or unconsciously, a model belonging to one story to insert it into another. Thus, in the Pierpont Morgan Collection (Metropolitan Museum of Art, inv. 17.190.1625), G. Gamot's watch is decorated on its lid with a *Judgment of Paris* and on the bottom with *The Partings of Aeneas and Dido*, after Bourdon, an episode interpreted no doubt as a *Rape of Helen*. The writer of the catalogue believes that the scene is taken from the story of Antony and Cleopatra.[64] The similarity of these scenes can easily lead to confusion indeed when iconographic references are lacking.

108a

108b

109

108a,b. Round watch. Painted enamel gold case and dial. Mid-seventeenth century. The movement is at the back. D. 65; Th. 22.
The case bottom (a) is ornamented with *The Judgment of Paris.* The lid (b) represents *The Rape of Helen.* The rich colors, the perfection of the drawing and the delicate workmanship make this a masterpiece of enameling.
Paris, Louvre, Olivier Coll., inv. OA 8317

109 Round watch. Painted enamel gold case. Movement signed "Louis Baronneau A Paris." Mid-seventeenth century. D. 46.5.
The case and the dial are painted in a grisaille on enamel. They represent battle scenes in the style of A. Tempesta.
Toulouse. Musée Paul Dupuy, inv. 18230

Cases representing couples from contemporary history in an allegorical form were sometimes commissioned from the enamelers of Louis XIV's reign. The most beautiful evidence of this kind is the case made by Henry Toutin in commemoration of the wedding of William of Orange and Mary Stuart, which we had the opportunity to describe earlier. A watch whose case was made by Pierre Signac, a painter on enamel for the queen of Sweden, now in the Historical Museum in Stockholm, is related to H. Toutin's masterpiece through its baroquely inspired allegorical scenes. On one of the outside faces, the queen driving a chariot receives the triumph of arms, while, on the other, Neptune offers her his scepter; inside, Queen Christina and her favorite are disguised as Atalanta and Meleager.

Except for the famous lovers of Fable and History, battles borrowed from ancient history most often decorated watchcases. In this case they often found inspiration in *The Battle of Constantine* by Jules Romain (Vatican), *The Passage of the Granicum*, a part of the *History of Alexander* painted by Charles Lebrun, and battle scenes engraved by Antonio Tempesta (1555–1630). A work from this master, for instance, decorates the counterenamel of a watch of the Olivier Collection (inv. OA 8431). A watch by Louis Baronneau is ornamented with paintings made in the same style (Paul Dupuy Museum, inv 18230).

Scenes out of the Old and New Testaments also decorated watchcases with success. Popularized through prints, the *Holy Family* and *Virgin with Child* works of Simon Vouet were often used by enamelers. A *Virgin with Child* of this painter's, engraved by Michel Lasne, served as a model for painting the outside of a watch lid by Josias Jolly in the Olivier Collection (inv. OA 8428). The bottom of the case is ornamented with a scene representing *Virgin and Child in the Company of Saint John the Baptist and Saint Elizabeth*. Its source is an engraving by Petrus de Jode II after Erasmus Quellin II (1607–1678).

Many preserved cases show analogies in style and technique, but it is difficult to attribute their production to a same enameler or a same workshop, not only because the degree of improvement in technique when these pieces were made can explain the resemblances, but because enamelers habitually used the same models, engraved to their specifications by designers, or reproduced painted works.

110 *Virgin with Child*, painting by Simon Vouet (1590–1649), engraved by Michel Lasne.
Paris, Bibliothèque Nationale, Cabinet des Estampes

111 Round watch. Painted enamel gold case. Movement signed "Jolly A Paris." Around 1640. D. 58; Th. 18.
The outside of the lid is painted after *Virgin with Child* by Vouet engraved by Lasne. The solidly graded relief of the shapes, the frank contrasting colors and the importance given to the figures closely relates this piece to these decorating the watches of Baronneau (ill. 107), Chevallier (ill. 112), Goullons (ill. 113) and Bouquet (ill. 114).
Paris, Louvre, Olivier Coll., inv. OA 8428

111

112 Round watch. Painted enamel gold case. Movement signed "R Chevallier A Blois." Around 1630–1635. D. 50, Th. 16.
The case bottom is ornamented with *The Rape of Amphitrite by Neptune*; the lid represents *Perseus Rescuing Andromeda.* This piece offers precious evidence because its movement, made by a watchmaker who died in 1636, places it at the beginning of painting on enamel.
Paris, Louvre, Olivier Coll., inv. OA 8429

113a,b. Round watch. Painted enamel gold case. Movement signed "Goullons A Paris." Mid-seventeenth century. D. 61; Th. 22.
The case lid reproduces a *Virgin with Child* engraved in 1638 by Claude Mellan, after a Vouet painting. The bottom represents a *Holy Family* by the same painter, dated 1640, known in the workshops through a Pierre Daret engraving.
La Chaux-de-Fonds, Musée International d'Horlogerie, inv. no. 554

◁ 114 Round watch. Painted enamel gold case. Movement signed "David Bouquet Londini." Mid-seventeenth century. D. 64; Th. 24.
The piece is very close in its style and subject to the Baronneau watch (ill. 107). On the bottom, *Venus Attended by the Three Graces* appears, and on the lid the goddess seems to want to disarm her lover Adonis.
Milan, Museo Poldi Pezzoli, inv. 3446

115 Round watch. Painted enamel gold case. Movement signed "Goullons A Paris." Mid-seventeenth century. D. 61; Th. 22.
Decorated on the outside with paintings reproducing works by Vouet, the watch is decorated on the inside with landscapes in the style of Henri Mauperché (1606–1686), a renowned Parisian landscape artist. Notice, on the same watch, how enameling workmanship can adapt to its model. In religious scenes the colors are boldly contrasted and the composition is broad. In landscapes, the organization is in small details and the colors are subtly graduated.
La Chaux-de-Fonds, Musée International d'Horlogerie, inv. 551

A principal group of watchcases made around 1630–1640, can be distinguished, characterized by scenes with few characters, three or four at the most. Seen from the waist up, the figures occupy almost the whole surface of the painting. They stand out on the white enamel background, which is barely covered with paint. All interest centers on the action shown; scenery is in a sense absent. The figures are clearly characterized by wide, oval faces. The treatment is vigorous. The firm design of the bodies corresponds to the bold oppositions between the colored surfaces. Very small, steady touches produce, in the eye of the observer, a fluid and glossy effect.

A "basin watch" in the Olivier Collection (inv. OA 8429) that can be placed with certainty within the years 1630–1635 and is signed by the Blesian watchmaker Robert Chevalier has a case whose outside faces are decorated in that style, with the *Rape of Amphitrite by Neptune* and *Perseus Rescuing Andromeda*. The previously mentioned watch of the Olivier Collection, whose movement is attributed to Josias Jolly (inv. OA 8428), presents a case painted, in the same manner, with religious scenes.

The firmness of the drawing, the contrast of the colors, made conspicuous by the light ground, and the importance given to the figures also characterize a watch by Goullons preserved in the Musée International d'Horlogerie (inv. 554), decorated with religious scenes painted after Simon Vouet. The watch of this museum, decorated with scenes from the Love of Venus and Adonis, already mentioned, is painted in that style, as is also the watch in the Poldi Pezzoli Museum in Milan, decorated with scenes certainly inspired by the same series of models (inv. 3446).

A pictorial style completely different from the preceding is found on cases enameled during the same period. It is a miniature style which evokes illuminations. The characters are presented in a very small scale and are dominated by the scenery around them. The composition is free, made with touches shaped like commas and uneven circles.

One of the most beautiful examples representing that manner is the case showing *The Construction of the Tower of Babel*, and containing a movement by M. Wentzel, watchmaker in Strasbourg (Musée International d'Horlogerie, inv. 1331).[65] The inside of the lid and cuvette of watches signed Goullons and Chevallier is decorated in this style, with scenes of daily life in the time of Louis XIII, while the outside is painted in a broad and vigorous manner with historical scenes. The juxtaposition on the same case of two different techniques and inspirations should warn us against hasty attributions of a particular style to a particular workshop.

A third group is composed of watches made between 1630 and 1660, presenting a highly perfected technique capable of rendering all the nuances of the models used by enamelers. In this series belong the "basin watches" decorated with episodes from the History of Theagenes and Chariclea, the Love of Aeneas and Dido and the Trojan War and the "historical watches" of H. Toutin and P. Signac. The watch in the Olivier Collection signed by Elias Weckherlin (inv. OA 8320), decorated on the bottom with *The Meeting of Aeneas and Dido* and on the lid with *The Parting of Aeneas and Dido* after S. Bourdon, is a magnificent example of the enamelers' skill in that period.[66]

Stippling offered an infinite array of nuances for the treatment of bright colors in drapes and gave the modeling of flesh an extreme delicacy. When the band and inside surfaces were not painted with scenes, they were often ornamented with landscapes animated with small, picturesque figures among ruins, copied from the works of Italianized landscapists of the seventeenth century, such as Henri Mauperché (1606–1686) and Gabriel Pérelle (1603–1677).

This series of watches, which reveals an admirable technique, served, in our view, as an example to the Huaud brothers; we find here the essential elements of their manner: the treatment by little touches, the colors that will characterize their palette (yellow, red, olive green and blue), the landscapes that will decorate the counterenamels and the bands of their cases.

Following the example of Petitot, most talented enamel painters will prefer to devote themselves, starting with the reign of Louis XIV, to the portrait genre. The close alliance of this genre and painting on enamel lets us suppose that the watchcases decorated with portraits might not have been rare in the seventeenth century and that they even constituted an important demand from a clientele responsive to the fashion of pendants and medallions containing the faces of loved ones. Isaac Gribelin, an enameler working for all the watchmakers in Blois, had indeed acquired his reputation in the art of painting portraits; we can safely say that he produced more than one watch thus decorated.

Two extant watches ornamented with royal portraits are evidence of the fashion. One, in the Victoria and Albert Museum, has a case painted, about 1640, inside the cuvette and the cap, with the portraits of Louis XIII and Richelieu (inv. 7543–1861). The other, in the Musée International d'Horlogerie, is decorated on the outside faces of the case with the portraits of Philip IV of Spain and Maria Anna of Austria, executed after Velazquez (inv. 1120).

116 Round watch. Painted enamel gold case. Movement signed by Goullons. Paris, around 1640.
Outside, the two lids that form the case are decorated with paintings figuring *The Holy Family* and *Virgin with Child.* They are ornamented on the outside with half-length portraits of Louis XIII and Cardinal Richelieu.
London, Victoria and Albert Museum, inv. 7543–1861

117

117 Round watch. Painted enamel gold case. Geneva (?), around 1675–1680. The movement is in the back. D. 33; Th. 17.
The half-length portrait of a plume-helmeted woman and four red camaieu landscapes ornament the case bottom. The work is very close to a watch in the Louvre (inv. OA 8327) signed by the Genevan enameler Jean-Pierre Huaud.
Paris, Louvre, Olivier Coll., inv. OA 8328

118 a,b. Round watch. Painted enamel gold case. Movement signed "Pierre Duhamel." Geneva, third quarter of the seventeenth century. D. 30; Th. 15.
The portrait of two young people vowing eternal love decorates the bottom. The paintings are remarkable for the meticulous workmanship, the graduated relief of the faces and the brightness of the colors.
Paris, Louvre, Olivier Coll., inv. OA 8433

118a

118b

We shall cnter into the category of cases decorated with portraits a few pieces belonging to a group of watches dating from the last third of the seventeenth century that have the common characteristic of being decorated on the bottom or on the inside of the cuvette with apparently similar medallion portraits. These portraits represent, half-length, in a three-quarter profile, a young woman in a plumed helmet.

In some portraits, like the one signed "D. Champion" (Olivier Collection, inv. OA 8333), she wears a cuirass, and in this case she appears to be the war goddess Minerva. In others, as on the watch by Jean de Choudens, also in the Olivier Collection (inv. OA 8316), she wears a low-cut dress. It is easy to notice that the faces and the busts are not uniform. There is also a noticeable difference between the juvenile figure decorating the watch we just mentioned and the one on another watch in the Olivier Collection (inv. OA 8328), of a heavier type. It is possible, we think, that these half-length portraits represent mortals who, through coquetry and a taste for fancy-dress, are decked out in the elegant headdress of Minerva and prefer to put on a gown enhancing their charms.[67]

The genuine double portrait of young people in mid-seventeenth century dress, promising each other eternal love, decorates the bottom of Pierre Duhamel's watch belonging to the Olivier Collection (inv. 8433). The counterenamel decorated with the bust of a young woman in contemporary costume and a plumed helmet relates the piece to the series we just mentioned. It is not unlikely that these watches may have come out of the same workshop, specializing in the portrait genre in contemporary costume or fancy-dress. This hypothesis is supported by the fact that several watches with the face of Minerva contain a movement by Pierre Duhamel: one cannot see why the watchmaker would have turned to several different workshops for cases decorated in the same way. Thus, the plumed figure of the Duhamel watch in the Victoria and Albert Museum is framed with a flower garland in a naturalistic style, similar to the one that surrounds the portrait of the Olivier watch.

The example of the watch by Duhamel in the Olivier Collection, presenting two different types of portraits, is not unique. Indeed, it is related to another watch in the collection, also decorated with a real portrait on the bottom and with a helmeted figure on the counterenamel (inv. OA 8327). This watch, whose movement is signed by a watchmaker named Christianus, interestingly bears the signature of the enameler Jean-Pierre Huaud. It thus provides proof that the workshop of the Huaud brothers practiced that kind of decoration. This is, perhaps, the first evidence that allows us to attribute enameled cases in this style to the Huauds and their shop.

Decoration of the movements

From the end of the seventeenth century, the cock became the most carefully ornamented element of the movement.

Fixed on the plate by a foot, it first presented a serpentine shape that soon became a foliage arabesque. Then it covered the annulary foliot entirely, while the foot widened.

From the first years of the seventeenth century until the appearance of watches with the regulating spiral spring, the part covering the balance wheel always had a more or less elongated ovoid shape. However, the foot presented various contours, going from rectangle to oval. The whole was in open work, engraved with flowery foliage, among which strawberry flowers were dominant. This decoration was repeated on the pawl, another part of the movement.

At times, other ornaments gave the movement a refined aspect. The watchmaker's signature was underlined with a palm, a floret or a flowery branch. The barrel and the pierced plate protecting the counting wheel of the striker were decorated with engraved foliage.

In that period, more or less tall, pear-shaped pillars were the most current. Molded baluster-shaped pillars were also often used. Marin Etienne offered several carefully drawn models in his manuscript.[68]

II Decoration from 1675 to 1720: appearance of national styles

The fantasy, variety and luxury that characterized watch decoration until the mid-seventeenth century were abandoned during that period, to be replaced by well-bred moderation and even a certain austerity. Watches were no longer considered jewels, but instruments with a practical utility. The invention in 1675 of the regulating spring spiral had changed them into true time-keepers.

Louis XIV watches: the so-called *oignons*

The introduction of the hairspring in watch movements provoked the appearance in France of a type of round watch, easily recognized by its clearly defined and constant characteristics, the pro-

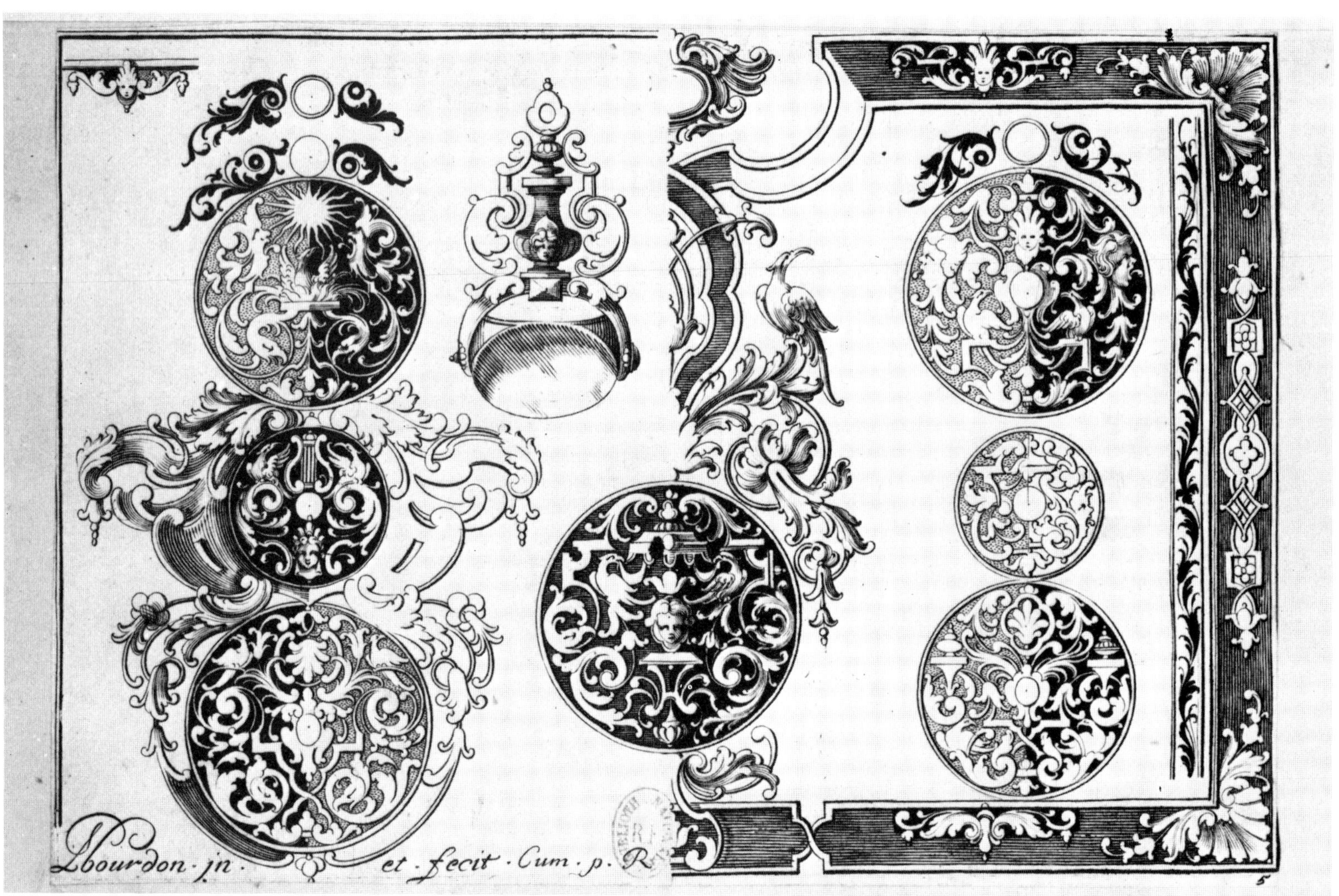

120 *Oignon* watch. Engraved gold case. Movement signed and dated "Franciscus Josephus De Camus Invenit An 1709." Paris. D. 62; Th. 46. Exceptional for its gold case and its movement, equipped with both "on demand" and *au passage* striking train, described by Camus in 1722 in his *Traité des forces mouvantes* (Treatise on moving forces), the watch belonged to Pajot d'Ons-en-Bray (1678–1754), a postmaster general and member of the Académie Royale des Sciences, who owned a famous cabinet of curiosities.
Paris, Louvre, Olivier Coll., inv. OA 8310

◁
119 Models for watchcases and cocks engraved by Pierre Bourdon. Paris, around 1703.
Paris, Bibliothèque Nationale. Cabinet des Estampes (le 49 rés., p 25)

duction of which lasted till the end of Louis XIV's reign. Its bulging and ample shape gave it the name *oignon*, or "onion," as it has since been known (in England, it was and is the "turnip").

The *oignons* still in existence, have either gilt brass or silver cases, left bare or engraved with ornaments, or metal cases covered with sharkskin or studded leather. The extreme scarcity of gold *oignons* can be partly explained by the Sumptuary Edicts promulgated against luxury at the end of the seventeenth century and also by the melting of a large number of them, for because of their size, they represented a rather important quantity of precious metal. The watch in the Olivier Collection made in 1709 by the watchmaker F. -J. De Camus for Count d'Ons-en-Bray, offers the exceptional example of a gold case (inv. OA 8310).

Watchcases engraved in the style of Bérain, Bourdon, Bourguet, Marot, etc.

Engraving and chasing are the processes most used in the decoration of metal cases. Taking their inspiration from models proposed

121 Models out of the *Livre de Boîtes de Pendules, de Coqs et Estuys de Montres et autres nécessaires aux Orlogeurs* (Book of Pendulum Clock Cases, Cocks and Watchcases and other necessities of the horologists), engraved by Daniel Marot (1655–1718). The Hague, early eighteenth century.
Paris, Bibliothèque Nationale, Cabinet des Estampes, Ha 8, p 195

by ornamenters of the period, Jean Bérain, Pierre Bourdon, Briceau, Jean Bourguet and Daniel Marot, the goldsmiths cover their cases with compositions whose symmetry is balanced by the flexibility of the decorative lines and by the variety and fantasy of the ornaments. Perched birds, sphinxes, genies, grotesque masks, palms, shells, tiny mythological scenes and monogrammed cartouches appear among lambrequins and scrollwork with acanthus foliage.[69]

Pierre Bourdon, an engraver from Coulommiers, worked in Paris in the first years of the eighteenth century. His *Essais de Gravure. Où l'on voit de beaux contours d'ornements traités dans le goût de l'Art, propre aux Horloguers, Orfèvres* . . . (Essays on Engraving. Where beautiful outlines of ornaments treated in the style of the Art are seen, appropriate to Horologists, Goldsmiths . . .), published in 1703, gives drawings of cases, cocks, keys and hands in which large acanthus leaves forming symmetrical arabesques are predominant. Bourdon notably proposes decorations conceived for cases meant to contain striking movements. Their composition is always the same: cartouches, alternatively full or pierced, follow each other along the circumference of the case bottom, the center of which is occupied by a medallion.

The striker watch, both *au passage* and "on demand," that belonged to Count Pajot d'Ons-en-Bray has an engraved and repoussé case in the style of these models. The bottom circumference, pierced in parts, is ornamented with four cartouches representing the seasons, while the center is decorated with a medallion—a flying eagle, its head turned toward the sun—framed with the motto "RIEN. DE. BAS. NE. MARESTE" (Nothing low can stop me) and with a garland composed of draperies, palms, acanthus leaves, birds, lions and two winged genies. Even though the case reflects Bourdon's manner, it is first of all an original creation whose iconography was elaborated to exalt the glory and the noble qualities of its owner, a postmaster general and member of the Académie Royale des Sciences, summarized by the motto on the bottom. Few of the preserved watches present decorations whose execution was inspired by the very personality of the patron.

122 *Left. Oignon* watch. Case covered with black Morocco leather and gold studded. Movement signed "Marchand A Genève." Around 1700. D. 53; Th. 34.
Right. Oignon watch. Case covered with black Morocco leather and gold studded. Movement signed "Gribelin A Paris." Late seventeenth century. D. 59; Th. 44.
Studded leather was a decorative process much in vogue during the late seventeenth century and early eighteenth.
Paris, Louvre, Olivier Coll., inv. OA 8421 and OA 8307

123 *Left. Oignon* watch. Leather-covered case. Movement signed "Rabby A Paris." Early eighteenth century. D. 60; Th. 37.
The white enamel dial with embossed hours is a characteristic element of many *oignon* watches. One hand is missing.
Right. Oignon watch. Engraved and pierced silver case. Movement signed "Simon Lalemant A Blois." Early eighteenth century. D. 50; Th. 35.
A large circular cock, engraved and pierced with a symmetrical acanthus-leaf decoration in the style of Louis XIV ornamenters, characterizes the *oignon* watches, like this one (with an alarm) made by a Blesian watchmaker received as a master in 1701.
Paris, Musée National des Techniques (CNAM), inv. 21958 and 21989

122

123

Daniel Marot's models show an inspiration very close to that of P. Bourdon. This Parisian, born in 1655, left France in 1685 for England, then established himself in 1702 in the Hague, where he died in 1718. The compositions in his *Livre de Boîtes de Pendules, de Cogs et Estuys de Montres et autres nécessaires aux Orlogeurs* (Book of Pendulum Clock Cases, Cocks and Watchcases and other necessities of the Horologers), of a great decorative richness, systematically use the lambrequin and the large and flexible acanthus leaf. The German engraver Paul Decker (1677–1713) produced watch designs decorated with grotesques, showing the clear influence of Jean Bérain. Briceau, a master goldsmith in Paris, published plates for the decoration of watches in 1709. In his models, the scrollwork traces elegant and fine arabesques, the acanthus leaves are elongated with a delicacy contrasting with the vigor with which they are drawn, in the designs of Bourdon and Marot.

Leather-covered cases

Leather or sharkskin cases, often decorated with gold or silver studwork, were very much in fashion throughout the period that the *oignon* dominated the watch production. Nails of various sizes composed a rosace or a monogram in the center of the bottom, and on the band and the bezel, friezes of circles, lozenges, fleurons or arabesques.

Evidence of that fashion is found in the inventories of estates. At the time of his death in 1703, President Lambert de Thorigny owned "a watch with a silver case made by Martineau, with silver chain and shagreen casket studded with small silver nails," valued at forty livres. According to the inventory of his possessions drawn in 1711, Arnoul Chapart, a Paris bourgeois, had "a silver-case watch made in Paris by Gribelin, with a silver chain, in its silver-studded casket," worth thirty livres.

Monograms often drew their inspirations from the *Livre des Chiffres* published in 1680 by Charles Mavelot, which contained "in general, all the names and surnames alphabetically arranged."[70]

Dial artwork

Oignon dials offer various compositions. Two types of dial can be distinguished, each characteristic of its production. In one, the brass dial plate has on its circumference white enamel cartouches outlined by fillets engraved in the metal, on which are painted, in black or blue, the Roman numerals of the hours; a white enamel circle marked with vertical strokes and dots indicating the half and quarter hours surrounds the center of the plate, which is engraved with motives in the styles of the ornamenters of the time. In the other, the entire dial plate is covered with white enamel on which Roman numerals are painted, indicating the hours; these rounded reliefs are often surrounded by a black or blue painted line, which makes them look like cartouches. A circle dotted with small painted motives indicates the half hours and the quarters.[71]

White enameled dials with entirely smooth surfaces and their hour numerals painted, made their appearance in the first years of the eighteenth century. They would gain the favor of all the dial makers of the reigns of Louis XV and Louis XVI.

After 1700 approximately, the indication of minutes became widespread. Dials started to have a circle on their circumference, divided by bars marking the minutes and numerals written for every five minutes. The minute hand was shaped like a point, its end often molded, the hour hand was carved like a fleur-de-lys, a fleuron or an arrow.

Decoration of the movement

Generally, the *oignon* movement is carefully decorated. The cock decoration is often painstakingly filled with fantasy and gives the plate a very refined aspect.

Whether gilt brass or silver, the cock in Louis XIV watches presents a large circular surface that permits the engravers to reproduce elaborate compositions inspired by collections of models proposed by ornamenters. The arabesques formed by acanthus leaves and lambrequins are often symmetrically arranged around a central motive—vase, grotesque mask or medallion. Crested birds, familiar or fantastic animals and cupids enliven the decoration. The lateral ears of the cock are ornamented with masks, acanthus leaves or shells.

Thanks to its large diameter, the cock was sometimes decorated with an enamel painted portrait. The portrait usually presents a half-length woman, her face seen in a three-quarter profile, her hair up and powdered, dressed in the fashion of the end of the eighteenth century. Though that face always conforms to the canons of feminine beauty then in vogue, one has the distinct impression that it is a true portrait. A watch in the Olivier Collection, signed by Jacques Gradelle, has a cock of that kind framed by a

124 *Oignon* watch. Painted enamel case. Paris or Geneva, around 1700. D. 48; Th. 35.
A portrait of a woman painted on enamel ornaments the cock, which lets a small steel heart, similar to a mock-pendulum, protrude. This kind of cock was received with some favor around 1700.
Paris, Louvre, Olivier Coll., inv. OA 8321

IACQVES GRADEL

125 Round watch. Silver case. Movement signed by Denis Miroglio. Geneva, early eighteenth century. D. 50.
The dial lets a succession of painted enamel figures parade past the window.
New York, Metropolitan Museum of Art, inv. 83. 1. 74

▷

126 Models of cases engraved by Simon Gribelin (1666–1733), intended for jewelers and watchmakers. London, 1697.
Paris, Bibliothèque Nationale, Cabinet des Estampes (Le 49 rés., p 4)

row of brilliants (inv. OA 8321). The example of a cock decorated with the portrait of a man is shown by a watch by Henri Debary, made around 1700, now in the Musée International d'Horlogerie. (inv. 566).

Watches ornamented with portraits—on the cock, but also on the dial or the case—appear to have gone through a particular vogue around 1700, not only in France but all over Europe. Like all portrait cases, they were very appreciated gifts.

Other parts of the movement were carefully decorated. Thus, the chain stop was often pierced and chased with symmetrical scrollwork. The pillars generally are little columns, square in section, split or not, and crowned with a capital. The model is known under the name of "Egyptian pillars."

Toward the end of Louis XIV's reign, the watch appears to have been conceived chiefly as a useful object. The robust look and lack of refinement of the majority of *oignons* contrast with the decorative affectation that was obvious in the earlier watches and would reappear in the Louis XV period.

But several innovations that appeared in the *oignons* will remain. From then on, the round case will dominate watch production. Its articulations are modified: the hinge that links the bezel to the cuvette is henceforth located on the level of the IX, the one of the movement is affixed above the XII, on a *bâte* (a raised mount) whose concave surface is topped by the bezel.

Swiss watches: the French influence

The installation of French watchmakers in Switzerland, the unceasingly growing sales of Swiss watches in France, the permanent relations established since the sixteenth century between the

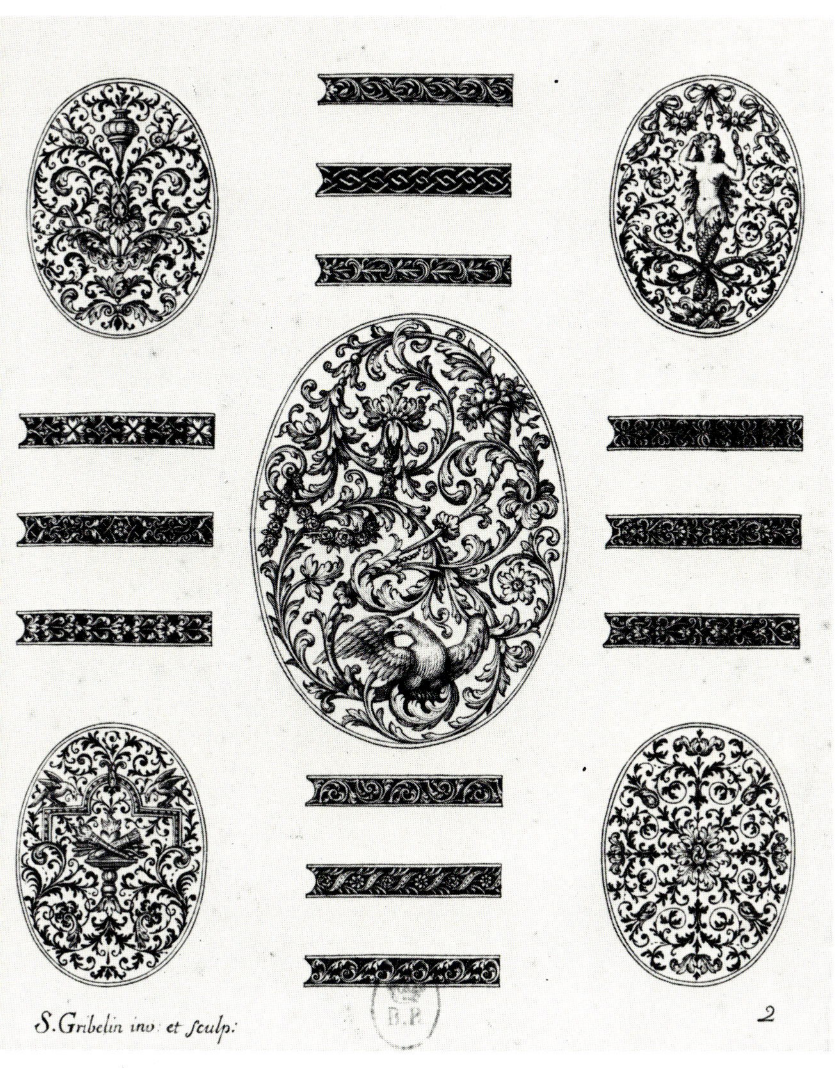

watchmakers of both countries, all contributed to strengthening the dominion of French fashions over the decoration of watches produced in Switzerland.

At the end of the seventeenth century, Swiss watchmakers often adopted the type of large, bulging watch that then characterized French watchmaking. Gilt silver or gold watchcases had a diameter of fifty-five to sixty millimeters and a thickness of about thirty-five millimeters. The metal is engraved and chased with decorative compositions in the style of Louis XIV ornamenters, or covered with chagreen leather, very often decorated with studs. A second case, called an *étui* (cover), sometimes protected the first, particularly when the latter was pierced in order to contain a striking train.

The white enameled dials are either "embossed" or smooth; the Roman numerals of the hours are elongated and painted with a steady hand. The metal dials present either white enameled cartouches or metal plates soldered to the surface; according to the case, the hour numerals are indicated in either painted or champlevé enamel. The center of these dials is often engraved with lambrequins and acanthus leaves that form a symmetrical design. Sometimes a window pierced on the dial lets one see a parade of half-length figures in painted enamel, representing characters from the contemporary theater. A watch having a dial of the kind, signed "Denis Miroglio, à Genève," is kept at the Metropolitan Museum of Art; a second one, bearing the signature of Baltazard Faure of Geneva, belongs to the Wilsdorf Collection.[72]

The decoration of the movement is closely related to the French model. The circular cock covers almost the entire plate, to which it is held by two screws sunk into two lateral ears. It is engraved and pierced with lambrequins, scrollwork with acanthus leaves, grotesque masks, shells, vases, etc. The watchmaker's signature is conspicuous, very often written in capital letters on the edge of the plate.

English watches

French and English watches did not show any essential differences until around 1680. At that time, while the *oignon*, which was to dominate French production for about forty years, was appearing, a typically English watch was created to exploit to the full the new technical inventions.

Decoration of pair cases

The circular shape was definitely adopted; the use of a double case, or "pair case," became systematic.

The interior cases are often made of gilt metal or silver. When intended for a nonstriking movement, they are generally undecorated. On the other hand, those with repeaters are engraved with pierced ornaments that let the sound of the bell be heard.

Cases of this kind are decorated around the circumference of the bottom with a chased openwork frieze. With a great decorative richness, it is composed of acanthus leaves forming large volutes, among which lambrequins, birds with spread wings and cupids appear. It is sometimes interrupted by medallions arranged symmetrically, unpierced and engraved with a landscape, a grotesque mask or an antique head. The central part of the bottom is either smooth, simply surrounded by an engraved border, or ornamented with a mythological scene, a landscape or grotesques.[73]

This decoration engraved in the baroque style was inspired by models proposed by French ornamenters of the Louis XIV reign, which the Huguenot émigrés, goldsmiths and watchmakers contributed to disseminate. Simon Gribelin's engravings were even

129 ▽ *127*△ *128*△

127 Round watch with an engraved and pierced silver pair case. Movement signed by Daniel Quare. London, early eighteenth century.
The outside case is pierced with a frieze of lambrequins and acanthus leaves, repeated on the inside case. The movement has a repeater striking train.
London, Victoria and Albert Museum, inv. 1362–1904

128 *Left.* Pair case round watch. Movement signed by James Markwick. London, around 1700.
The outside case, in silver pricked tortoise-shell, presents a decoration in the style of P. Bourdon.
Right. Pair case round watch. Movement signed "Johannes Bayes Londini." London, third quarter of the seventeenth century.
The outside case is in gold pricked tortoise-shell.
London, Victoria and Albert Museum, inv. 1148–1893 and 395–1888

129 *Left.* Pair case round watch. Movement signed by James Markwick. London, around 1700.
The silver dial has only one hand, indicating both the hours and the minutes.
Center. Pair case round watch. Anonymous movement. Dust cap signed by Gerret Bramer of Amsterdam. London, first quarter of the eighteenth century.
The white enamel dial, with two silver hands, presents a minutes circle, forming, in the Dutch style, an arcade succession.
Right. Pair case round watch. Movement signed "F Stamper London." End of the eighteenth century.
The hours of the day and night are indicated on a half-circle by two pointers shaped like the sun and the moon.
London, Victoria and Albert Museum, inv. 1148–1893, 654–1872, 46–1905

more of an influence on English goldsmiths when they were published in London itself in 1697. This engraver, born in 1666 into a famous Blesian family, immigrated around 1680 to the English capital, where he died in 1733. His collection, titled ***Book of Ornaments usefull to Jewellers, Watchmakers and all other Artists***, grouped models of various inspirations, with abundant use of acanthus.

The exterior cases of striking watches are often engraved with a decoration repeating the one inside. The Victoria and Albert Museum has a watch, its movement signed "D. Quare London" (inv. 1362–1904), whose two silver cases present, barring a few details, the same decoration pierced with acanthus, masks and birds.

Exterior cases were very often made of leather, sharkskin or tortoise-shell. When they were intended for protecting a striking movement, holes were pierced along their circumference. On leather and tortoise-shell, the fashion was most often a gold or silver studding.

A type of decoration known under the name of "pricked tortoise-shell" appears to be particular to English watches. By means of gold or silver inlays—according to the metal chosen in the fabrication of the inside case—the artist drew a naive landscape or delicate compositions formed with flowers, foliages or birds. Three such decorated watches belong to the Victoria and Albert Museum (inv. 1148–1893; 395–1888).

The English exterior cases, like the French, are articulated by a hinge located on the level of the IX. However, the interior cases have a hinge affixed above the XII, below the pendant. The shape of the latter is modified: the pendant with a ball-like head, through which the ring moved freely, intended for hanging the watch, gives way to a flat-headed, studlike pendant to which the watch ring, or bow, is fixed by two mobile screws. The watchmaker's term *bow* suggests the shape of this ring: an overcocked bow.

Variety of dials

A variety of dials characterizes English manufacture during the period 1680–1710. This diversity reveals the efforts of English watchmakers to work out the most legible type of dial indicating the hours and minutes.

The most common type and also the simplest, is the metal dial: rough grained, bearing small plates on which Roman numerals for the hours and Arabic numbers for the minutes, counted by fives and including the quarter and half hours, are indicated in champlevé enamel. A cartouche containing the watchmaker's name is often applied in the center of the dial.

Four other dial models are characteristic of this period of research.[74]

The "moon and sun" dial is fairly common; it reflects the taste of clients for the astronomical indications found on watches dating from the sixteenth century. The numerals for the minutes and their divisions are indicated along the circumference of the dial. Inside the circle, the half-circle formed by the hours numerals frames the semicircular window in which the sun and the moon appear. The hours are numbered on the right of the half-circle from I to VI and on the left from VI to XII; the sun and the moon, alternately serve as pointers to indicate the hour. Thus, when the moon has reached VI in the morning, it disappears under the plate of the dial to make room for the sun, which appears in front of the opposite VI. The bottom part of the dial may be occupied by the cartouche indicating the watchmaker's name and by a small window showing the date. In his ***Remarques qui pourront être de quelque utilité dans le choix des montres*** (Observations that can be of some use in the choice of watches) Savary classified this kind of dial among the "baubles": "It is . . . through

130 Pair case round watch. Movement signed by Johannes Will. Heidelberg, first half of the eighteenth century. D. 42.
Decorated with extreme refinement, the movement shows an obvious English influence.
Kassel, Hessisches Landesmuseum, inv. U 111

ridiculous inventions that Masters of a mediocre merit try to dazzle the eyes of those who do not have much knowledge. Is it not . . . a great curiosity to see on a Watch the Sun & the Moon always go up & down at 6?"[75]

The "wandering hour" dial lets the precise time be known at a glance by doing away with the hands. The numeral corresponding to the hour moves along a half-circle indicating the minutes. When, for instance, III is in front of the thirtieth division of that half-circle, 3:30 is indicated.

The dial known as the "differential dial" is much less common. The minute hand and the indication of minutes are placed normally. The hour hand, however, has been eliminated. In the center of the dial, a mobile disk carries the hour numerals. The disk rotates in such a way that the numeral corresponding to the hour lies under the minute hand. If the latter, for instance, covers the V and indicates the tenth division of the minutes, it is 5:10.

The dial may also have been divided into only six hours, so that a single hand would make its revolution twice in twelve hours. In this case, the Roman numerals I to VI were superimposed on the Arabic numbers 7 to 12. In each of the six intervals thus defined, the divisions of the minutes 0 to 60 were repeated. According to the interval it was moving in, the minute hand marked the hour. For instance, if the minute hand was located in the period between 7 and 8 and pointed at the twentieth division, it was 7:20.

These variants show that the watchmakers and their customers were not entirely convinced that the principle of two hands was the best. Perhaps they found it difficult instinctively to read the time from two hands.

The tulip-shaped hour hand was routinely used until the end of the seventeenth century, when it was replaced by a fleuron-shaped hand, called the *beetle* in England. The minute hand first had the appearance of a long, thin pointer; then it swelled and became sculptured. This shape, called the *poker*, lasted until the start of the nineteenth century.

Decoration of the movement

English watchmakers kept to the one-footed cock, but they broadened the disk so that it covered the whole balance. The foot, at first rather small, took the form of a portion of a circle and was parallel to the plate. The foot and the disk were richly decorated with chased, pierced foliage. Another plate, ornamented in the same manner, sometimes covered the rest of the plate protecting the movement. The heel was often engraved with a grotesque mask or a seashell. Toward the beginning of the eighteenth century, two ears, shaped like a bird's head or acanthus leaves, were added on either sides of the heel.

The pillar structure was rather varied; the so-called Egyptian model was the one of those most often used, along with the tulip, which survived into the mid-seventeenth century, and the one shaped like a baluster.

131 Pair case round watch. *Repoussé* silver exterior case by Cochin. Movement signed "C Uyterweer Rotterdam." Early eighteenth century. D. 52.
The movement has a mock-pendulum balance, protected by a cock engraved with the figure of a child brandishing a cross and facing a serpent, with the motto "Malgré l'Envie" (Despite Envy).
Milan, Museo Poldi Pezzoli, inv. DT 5

German and Dutch watches: the English influence

Watches produced in Germany and Holland in the end of the seventeenth and the beginning of the eighteenth century generally owe their shapes and decorations to contemporary English designs. A few details in dial and movement decoration, however, characterize the Dutch watchmakers' production.

Dials fashionable in England are frequently used by German and Dutch watchmakers, notably the "sun and moon" and the "wandering hour" models. A dial characterized by a minutes circle, designed as a succession of arcades, each corresponding to a five-minute interval, became particularly popular in the Netherlands.

The Dutch and German watches usually have an English type of cock. Engraved with foliage, the pierced disk and foot are joined by an unpierced heel reinforced by two ears that are often acanthus shaped. An openwork engraved plate covers the rest of the movement plate, enhancing its decorative richness.

The "mock pendulum" balances are rather common in Dutch watches. They might even be considered a peculiarity of Dutch

132 Round watch. Silver case. Movement signed "H Van Loon Haerlem." Haarlem, end of the seventeenth century. D. 43; Th. 25.
The movement, with a verge escapement and a hair spring balance, is characterized by four pillars shaped like standing lions.
Paris, Louvre, Olivier Coll., inv. OA 8416

production. An opening on the cock disk lets one see the back-and-forth movement of a circular metal plaquette fixed on one arm of the balance. The remaining part of the cock is at times chased with a figure or a scene accompanied by maxims. Several surviving watches by Cornelius Uyterweer, a watchmaker in Rotterdam, have cocks of the kind, bearing moral aphorisms.[76]

The Dutch watches have pillars like those used in England and in France (the so-called Egyptian and the tulip-shaped pillars), but they also have more original pillars, ornamented with silver or blued-steel plates, engraved and chased with foliage. A watch in the Olivier Collection, signed by Hendrik Van Loon in Haarlem, has pillars in the shape of upright lions.

An international vogue: the production of the Huaud brothers, enamel painters

During the last third of the seventeenth and the first quarter of the eighteenth century, the very talented Huaud brothers kept up the tradition of painting on enamel. The relatively high number of their cases still in existence lets us imagine the considerable importance of their production. The diversity of signatures on the movements contained in these cases is proof of their renown among watchmakers in all the European countries.

The activity of the Huauds was not limited to watch decoration, but little evidence reached us as to their other occupations.[77] A few portraits bearing their signature have been preserved. The half-length portrait of a *Young Lady*, in the Musée d'Art et d'Histoire in Geneva, is dated 1688 and signed by Pierre Huaud II. Two male portraits in the David-Weill Collection; a portrait of *Professeur Pictet* (Louis Pictet Collection); *Frederick William as a Child* (formerly in the Hohenzollernmuseum); a *Prince in His Cuirass* (Musée d'Art et d'Histoire, Geneva); and a portrait of a woman, in the Landesmuseum in Zurich, present the double signature of Jean-Pierre and Amy Huaud. Because of the quality of the drawing and the colors, the masterpiece of Jean-Pierre and Amy is a cup and its matched saucer owned by the Victoria and Albert Museum (inv. 957 and 957 A–1882). Both of these entirely enameled pieces are ornamented with medallions showing pairs of lovers: Venus and Adonis, Acis and Galatea, Vertumnus and Pomona, Flora and Zephyr. The acanthus foliages and the grotesque masks that separate the scenes are similar to those decorating the band of their watchcases. Jean-Pierre and Amy also painted miniatures on ivory and parchment. The inventory made after the death of Jean-Pierre in 1723 mentions several.[78]

Pierre Huaud, born in 1612 in Châtellerault into a Protestant family, emigrated to Geneva, where he was received a *habitant* in 1630. His goldsmith apprenticeship in Laurent Légaré's shop ended in 1634, when he became a journeyman. A little later, he obtained his mastership. On June 13, 1643, he married the daughter of a lapidary, Françoise Mussard, who gave him three sons: Pierre II, born February 2, 1647, Jean-Pierre, born July 28, 1655, and Amy, born August 9, 1657. Pierre Huaud, along with his sons, was received as a *bourgeois* in 1671.[79]

To his activity as a goldsmith, Pierre I joined enamel painting, as shown in a deed of indenture undertaken in 1661 with Jean André, in which he promised to teach "*la peinture en esmail*".[80] Pierre I was able to teach the technique to his own children before his death on January 4, 1680.

The only known work that could be attributed to him is a case, signed and dated "P Huaud P a G 1672," depicting *Diana and her Nymphs Surprised at her Bath by Acteon*, now in the Louvre (Garnier Collection, inv. OA 7077). But the piece may also be the work of his son Pierre, then twenty-five years old. At first, Pierre II did not emphasize in his signature his position as the first born, since his brothers, were eighteen and sixteen in 1672, and still must have been apprentices.

On October 20, 1678, Pierre II married Eve Delarue, with whom he had four children, baptized in Geneva, from 1679 to 1684. H. Clouzot thinks that he went to Berlin in 1685. The Great Elector of Brandenburg-Prussia was attracting to his State thousands of foreigners to repopulate it and enhance its economic development. In 1686 Pierre was in Geneva, while his two young brothers were appointed, perhaps through his effort, painter-enamelers to the electoral prince. Having left for Germany at the end of 1689, he was named painter-miniaturist of Frederick III, on September 19, 1691. He was still at the Court of the Elector in 1693, as attested by a petition that he addressed in favor of a Genevan goldsmith, Pierre Chastel,[81] a "maker of boxes and cases for watches." His official activity continued for the next few years, since a portrait of Elector Frederick III, who was to become Frederick I, king of Prussia in 1700, formerly in the Hohenzollernmuseum in Berlin, bears the mention "P. Huaud l'aisné [first born], pinxit Borolinae, 1696".[82] He died between that date and 1698, when his widow was mentioned among the Protestant refugees in Berlin.

Watchcases painted by Pierre Huaud II are, most often, signed in a cartouche on the band. He almost always mentioned his status as first born: "P. Huaud l'aisne pin a G" (Olivier Collection, inv. OA 8324), "Huaud l'aisné pinxit à Genève" (Musée du Petit-Palais, Tuck Collection no. 240).

The suppleness of the drawing, the delicacy in relief reproduction—obtained through a stippling technique—the brightness of his colors characterize his best works. Outstanding among them is a case in the Olivier Collection (inv. OA 8324) the bottom of which represents *The Embarkation of Chariclea*, after Charles Poerson's painting, in the Louvre. The enameler used warm and deep colors—orange, yellow, and red—that he boldly opposed to olive-green, a blue and a violet. The very delicate variations in the treatment, through small, juxtaposed touches, softens the violence of the contrasting colors and gives the flesh a very soft relief. The landscape on the counterenamel is carefully composed, with trees on the right and houses on the left and in the background. A very successful harmony of blue, green and orange-yellow is repeated, in a darker mode, in the clothing of the two men conversing in the foreground.

The same science of color and the same delicacy in the relief distinguish the *Mystical Marriage of Saint Catherine*, signed ". . . Aisné pinxit à Genève 1681," decorating a watch kept in the Musée du Petit-Palais (Tuck Collection, no. 231), whose model is a work by Nicolas Loir (1623–1679). The case decorated with *The Judgment of Paris*, in the Musée de l'Horlogerie in Geneva (inv. E 428), is also a masterpiece because of the easy naturalness with which the enameler was able to arrange the figures on the circular surface of the bottom, the unconstrained attitudes and the subtle gradations of the flesh tints in his nudes, whose pearly white luster is brought out by the red and the blue of the two drapings. These same qualities are found in a less ambitious scene with two figures, *Venus and Adonis*, signed "P. Huaud," decorating a watch in the Olivier Collection (inv. OA 8325). The bright hues of the colors—olive green, violet, blue, orange-yellow and red—enhances the whiteness of the goddess' naked body, sprawling in a languorous abandon.

Pierre Huaud's two younger brothers also had brilliant careers as enamelers, attested by a few archival documents and several dozen watches. Upon their father's death in 1680, Jean-Pierre, then twenty-five years old, was appointed Amy's guardian. Two years later, Amy, now reaching his majority, entered a partnership contract with his brother: "[T]he said MM. Jean-Pierre and Amy Huaud . . . freely and willingly declare that they associate for the purpose of working on enamel for the time and term of six entire and continuous years. . . . "[83] During the period, in order to mark the commonality of their work and to differentiate themselves from their brother Pierre, they generally signed their pieces: "the two Huaud brothers, the younger."[84] In November 1684 Jean-Pierre married a young woman born into a noble family of Poitou, Adrienne de Tudert, while around the same time, in a second wedding, Amy married Antoinette Dutoit, a minister's daughter.[85]

On May 18, 1686, the two brothers were appointed painter-enamelers of the elector of Brandenburg. Geneva was then looking with some misgivings at the expatriation of its craftsmen, which could cause the implantation of competing industries abroad. So the electoral prince wrote, in the name of his father, Frederick William, the Great Elector, a letter to the Council of Geneva to ease the departure of the young Huauds.[86] They were able to establish themselves in Berlin, where each received two hundred thalers as a yearly stipend.[87] The works they executed at that time are usually signed "les deux freres Huaud, peintres de son A. E. de B. à Berlin," that is, "painters of his Electoral Highness [*Altesse électorale*] of Brandenburg". In 1699 Jean-Pierre and Amy were mentioned along with their children in the list of refugees in Köln-an-der-Spree,[88] the Berlin section reserved to foreigners. The death

133 *Top.* Round watch. Painted enamel gold case. Movement signed by J. B. Baillon. Paris, around 1740. D. 50; Th. 22.
The case bottom shows *The Rape of Europa* in a rococo-style frame.
Left. Round watch. Painted enamel gold case signed "Huaud l'aisné pinxit à Genève." Movement signed "Lucas Amsterdam." Last quarter of the seventeenth century. D. 41; Th. 24.
The bottom, painted by Pierre Huaud II (1647–c. 1697) represents a cavalry fight, after Philippe Wouverman.
Right. Round watch. Painted enamel gold case signed "Huaud le puisné fecit." The movement is in the back. End of the eighteenth century. D. 40; Th. 22.
On the case bottom, Jean-Pierre Huaud (1655–1723) painted a scene copied from a work by Vouet, *Juno and Iris.*
Paris, Musée du Petit-Palais, Tuck Coll., inv. 234, 240, 233

134 Round watch. Painted enamel gold case signed "fratres Huaut pinxerunt." Around 1700. D. 36; Th. 20.
A *Holy Family with Saint Ann and Saint John the Baptist,* after a Rubens painting, engraved in 1620 by Vorsterman, decorates the case painted by Jean-Pierre and Amy Huaud.
Paris, Louvre, Olivier Coll., inv. OA 8441

135 Round watch. Painted enamel gold case signed "Les freres Huaut." Movement signed "Augustinus Rumell." Berlin, about 1700. D. 42; Th. 28.
On the bottom *Roman Charity:* the Athenian citizen Cimon is comforted by his daughter, who suckles him in his prison. The contrasting color scale (red, blue, yellow, violet) that characterizes the Huauds' work is shown in this painting.
Paris, Louvre, Olivier Coll., inv. OA 8477

134△

133 135

136 *Mercury and the Three Graces*, engraving by Michel Dorigny, dated 1642, after a painting by Simon Vouet.
Paris, Bibliothèque Nationale, Cabinet des Estampes

▷

137 Round watch. Painted enamel gold case, signed "Huaut le Puisné." Movement signed "De Combe A Paris." End of the seventeenth century. D. 40.
Pierre Huaud copied Vouet's work, *Mercury and the Three Graces*, on the bottom of the case.
Milan, Museo Poldi Pezzoli, in. 716

of their older brother in Germany between 1696 and 1698 is certainly one of the deciding reasons for their return to Geneva in 1700.

Their shared activity went on during the whole first quarter of the eighteenth century, until Jean-Pierre's death on February 6, 1723, followed by Amy's on November 16, 1724.[89] The inventory of Jean-Pierre's estate shows, through the variety and abundance of the pieces mentioned, that their reputation had not suffered. One notices enamel portraits, enameled snuffboxes and watchcases, as well as miniature portraits and religious scenes. During that last period of their careers, the two brothers had certainly renewed the partnership act that had united them in 1682, for the work carries the double signature: "Les deux frere huaud l'ont faite" (Olivier Collection, inv. OA 84400), "Les deux frere Huaut fecit" (Olivier Collection, inv. OA 8435), "fratres Huaut pinxerunt" (Olivier Collection, inv. OA 8441).

Jean-Pierre, however, signed a few pieces alone. His most frequent signature is "Huaud le puisné fecit." He is responsible for pretty cases decorated after paintings by Simon Vouet, such as the one on a watch in the Musée du Petit-Palais (Tuck Collection, inv. no. 233) representing *Juno and Iris*, the one of a watch in the Musée de l'Horlogerie in Geneva, decorated with *The Rape of Europa* or the one of a watch of the Museo Poldi Pezzoli, representing *Mercury and the Three Graces*.

The enamel paintings bearing the signature of the Huaud brothers are always high quality. They are often inspired by the amorous adventures of the ancient gods and heroes, Venus and Adonis, Aurora and Cephalus, Diana and Endymion, Mars and Venus, Aeneas and Dido, Antony and Cleopatra, Meleager and Atalanta. But a fairly large number set the scene for Old Testament figures: Rebecca and Eliezer, Susanna and the Elders. Others illustrate the life of Christ: the Nativity, the Adoration of the Shepherds, the Holy Family. A subject frequently shown by the Huauds is the Roman Charity.

The Huauds seldom, if ever, created an original composition. They executed their paintings from those popularized through etchings, such as those by Peter-Paul Rubens, Simon Vouet and Laurent de La Hyre and later those of Antoine Coypel. Let us cite a few examples chosen in the Olivier collection: a case signed "Les deux frères Huaut fecit" is decorated by a Simon Vouet work, *Aurora and Cephalus*, engraved in 1642 by Dorigny (inv. OA 8435); a case signed "les deux frère huaud l'ont faite" is ornamented with a *Judgment of Paris* of Laurent de La Hyre (inv. OA 8440); a case signed "fratres Huaut pinxerunt" presents *The Holy Family with Saint Ann and Saint John the Baptist*, from a Rubens painting, engraved in 1620 by Vosterman (inv. OA 8441).

Thus, we have verified that the Huauds most often found their inspiration in works from the first half of the seventeenth century.

138 *The Judgment of Paris*, engraving by Antoine de Fer from a painting by Laurent de La Hyre (1606–1656).
Paris, Bibliothèque Nationale, Cabinet des Estampes

139 *Left.* Painted enamel gold case signed "Les deux frere Huaut fecit." Geneva, around 1700. D. 38; Th. 21.
A scene figuring *Aurora and Cephalus*, painted after a Vouet work engraved in 1642 by Dorigny, decorates the bottom.
Right. Round watch. Painted enamel gold case signed "Les deux frere huaud l'ont faite." Geneva, around 1700. The movement is in the back. D. 30; Th. 19.
The Judgment of Paris, after the work by Laurent de La Hyre, decorates the case bottom. The garnet-red, bright blue and orange-yellow so characteristic of the Huauds' work is found once again.
Paris, Louvre, Olivier Coll., inv. OA 8435 and OA 8440

140 *Left.* Round watch. Painted enamel gold case signed "Les deux frere Huaut les jeunes." Geneva, around 1682–1686. The movement is in the back. D. 37; Th. 22.
The Meeting of Eliezer and Rebecca decorates the bottom.
Right. Round watch. Painted enamel gold case signed "Les deux frere Huaut les jeune." Geneva, around 1682–1686. The movement is in the back. D. 39; Th. 22.
The Nativity, from the Rubens painting, engraved by Bolswert, decorates the bottom. The contrasting colors are softened by a skillful artistry, obtained with multiple gradated small touches.
Paris, Louvre, Olivier Coll., inv. OA 8439 and 8437

139

140

Their repertoire remained traditional, even during the Regency. A few works, made at the end of their career, however, are evidence of an effort to renew their inspirational themes. A cup and saucer in the Victoria and Albert Museum are decorated with *scènes galantes* inspired by paintings marked by the lighthearted spirit of Louis XV style. One of them, presenting *Flora and Zephyrus*, after Antoine Coypel, also decorates two watchcases in the Olivier Collection, one of them with the Huauds' signature. The refined subtlety of the colors harmonizes with the idyllic charm of the subject: golden yellow, garnet pink and cobalt blue shatter in multiple shades on the surface of draperies and are echoed, in a minor key, in the flowery garland surrounding the tenderly embracing couple.

Like their older brother, the two younger Huauds used a scale of warm colors made more exciting by cold tonalities. They almost always utilized a beautiful purple-red, an orange-yellow in opposition to a cobalt blue, a vivid green and a bluish purple. The surfaces in their paintings are broken up by small touches in graduated light and shades. The technique helped them obtain a softly blended relief.

Very beautiful examples of their production are part of the Olivier Collection. The same richness of colors and the same perfection of technique distinguish scenes as varied as *The Meeting of Rebecca and Eliezer* (inv. OA 8439), *The Rape of Cephalus* (inv. OA 8435) and *The Nativity* (inv. OA 8437) copied from a Rubens painting engraved by Bolswert.

The counterenamel and the band of the cases painted by the Huauds are generally decorated with landscapes inspired by the works of Henri Mauperché and Gabriel Pérelle.

The band is usually decorated with four oval medallions depicting landscapes with various buildings—towers, castles, cottages, ruins—often situated next to a river or pond. The warm tonalities of foliage and constructions are set off by the graduated blues of water, sky and hills. These medallions are separated by grotesques, fleurons or intercrossing knots, drawn either yellow on a brown or blue background or black on white.

The counterenamel usually shows a rolling countryside characterized by one or two tall trees on the foreground, a group of constructions, a watery surface and the discreet presence of a traveler. As in the landscapes on the circumference, the green and russet tonalities of the foliage harmonize with the bluish tones of distant hills and of the sky. The harmony is very delicately set off by the colors of the clothing of a small figure in the foreground. The Huauds often succeeded in giving these tiny landscapes with their carefully balanced masses a sunny lighting.

Many nameless cases unquestionably reflect the characteristic style of the Huaud brothers: the same stippling technique, the same scale of warm and deep colors, the same band and counterenamel decoration. These cases evidently either came out of their shop or were the work of skillful imitators.

A few watchcases have preserved the names of Genevan enamel painters who were active during the end of the seventeenth century and for the first years of the eighteenth. Thus, we recently saw, for sale, a case signed "Mussard pinxit" representing *The Mystic Marriage of Saint Catherine*, after Nicolas Loir, and by the same enameler, a case adorned with an elegantly erotic scene signed "J. mussart pinxit."

The Louvre preserves a watch decorated with a painting representing *Venus and Vulcan in the Company of Cupid*, signed and dated "Jean André pinxit 1687" (inv. OA 10 082). Another piece by that enameler, who was apprenticed in the shop of Pierre Huaud I, signed and dated "André pinxit 1716," was part of an auction sale in May 1984.[90] A watch decorated with a *Holy Family*, signed "Jean André pinxit," was shown in the Musée de l'Horlogerie in Geneva during the exhibition devoted to "Watches of Geneva, 1630–1720." The same exhibition contained a watch decorated with a *Judgment of Paris*, signed "J. L. Durant pinxit."[91]

Jean Mussart, Jean André and Jean-Louis Durant created works strongly inspired by the style of the Huauds, but of mediocre workmanship. The remarkable qualities characterizing the composition, the relief and the colors of the scenes painted on enamel made around the mid-seventeenth century—which still appear, already weakened in most of the works of the Huauds—have disappeared by the start of the eighteenth century.

III Decoration from 1730 to 1820

In the eighteenth century, watches were important accessories of feminine and masculine attire. To make them really conspicuous on the dress, it was very much the fashion to wear them hanging from the belt, on chatelaines.

The watchmakers had to accommodate rapidly and with ease the unceasing demands for novelty and produce watches decorated in the taste of the day. Because of their dependency on fashion, watches are a faithful reflection—like snuffboxes, to which they are related by their decoration and their quality as personal objects—of the evolution of styles. At the same time, they present the interest of revealing the various nuances by which these styles were interpreted, according to the countries in which they were decorated.

The French watch

The decoration of the cases intended for the French clientele—whether they were fabricated in Paris or Geneva—is always stamped with a certain elegance that may sometimes be overly pretty but is never too showy. The French watch of the eighteenth century and the first quarter of the nineteenth combines with refinement all the decorative techniques adapted to watchcases and other jewelry by seventeenth-century artists. Thus, they present the interest of being the only type of jewelry receiving enamel decoration in a period when jewelers had abandoned enamel, which was used so often in the sixteenth and eighteenth centuries. It also gives evidence to the persistent vogue of painting on enamel in France during the eighteenth century.

But few cases painted in the period are the work of creative artists, even if some of them appear of a superior quality and with an original inspiration, it would be hard to attribute them to illustrious enamel painters. The famous enamelers of the century, André Rouquet, Durand, Jacques Thouron, Hall, Pierre Pasquier and Nicolas-André Courtois, all preferred portraiture to the art of painting watchcases.

The years 1730–1760: the rococo style

Always circular, made in brass, silver or gold, embellished with engraved and chased motives and paintings on enamel, the cases painted under Louis XV all have the delicate prettiness of the rococo style of the period.

Decorative composition of enameled cases

The cases decorated with paintings on enamel were very much in demand by the clientele of the day. When they are enameled, "en plein" [in full], the painted enamel occupies the whole bottom of the cuvette or at least a large part; the bezel and the counterenamel in this case are also painted.

The tradition of making a counterenamel, inherited from seventeenth-century enamelers, progressively disappeared as painted enamel was used on a more limited surface on the bottom. A few of the Louis XV cases are counterenameled. The counterenamel, inspired from the Huauds' style, often represents a landscape with, for instance, a river shore, cottages or ruins. The composition is simple, even naive, and evidently cannot bear comparison with that of the inside paintings of seventeenth-century cases, which are real miniature pictures. The enameled bezels also are generally decorated with tiny landscapes animated with figures.

Besides the enameled scenes, the decorated cases offer large engraved and chased parts. One notes, however, that on Louis XV cases, the enameled scenes always take an important space on the bottom; in the following period that surface will be reduced to a small medallion. The frame of the paintings is then composed with engraved and repoussé motives, enhanced at times with opaque or translucent enamel, coming from the repertory of rococo shapes: volutes, seashells and flowery garlands. The same motives also decorate the bezels.

The enameled scenes form a harmonious composition with the engraved ornaments. The harmony is enhanced by the refinement of numerous details: the head of the pendant, the button and the catch of the case are often enriched with diamonds, the hands on the dial are at times ornamented with diamond rosettes. The hinge, seen at the level of IX, does not disturb the harmony of the whole; it is solid, yet very decorative, gracefully curved at the extremeties and engraved with fillets.

ENAMEL PAINTINGS IN THE STYLE OF WORKS OF FASHIONABLE PAINTERS

The enamel paintings that decorate cases of the period represent, for the most part, *scènes galantes*, often in sylvan surroundings, showing figures conversing, at play or giving a concert. Very often, they were inspired by engravings of the works of Watteau (1684–1721), Lancret (1690–1743), Pater (1695–1736), Carle Van Loo (1705–1765), Boucher (1703–1770) and Fragonard (1732–1806). The Olivier Collection has numerous watches enameled after these painters, as well as some after Lemoyne (1688–1737) and Cazes (1676–1754). We shall cite a few examples chosen from the Louvre collection and those in other museums.

A Watteau painting, *The Lesson in Love* (Stockholm, National Museum), decorates a watch in the Musée International d'Horlogerie (inv. no. 1114). Another work by Watteau, entitled *The Storyteller* (present location unknown), engraved in 1727 by C.-N. Cochin, is reproduced on a Julien Le Roy watchcase, now in the Olivier Collection (inv. OA 8340).

The allegory, *Music*, painted by Carle Van Loo for Bellevue Castle, which belonged to Marquise de Pompadour, and engraved in 1757 by Etienne Fessard, seems to have been copied often in enamelers' shops. The painting is now in San Francisco in the California Palace of the Legion of Honor. It notably decorates a Pierre Le Roy watch in the Olivier Collection (inv. OA 8345), another by Baillon in the Musée Paul Dupuy (inv. 18258) and one signed "Caron à Paris," auctioned at the Hotel Drouot on April 1, 1938. The work also decorates a snuffbox in the Wallace Collection (inv. no. XVIII-A 86). The connection between watch and snuffbox decoration is constant throughout the eighteenth century.

We have noticed, for instance, three watches decorated after Fragonard's *Blindman's Buff* (Toledo, Museum of Art), painted around 1755 and engraved by Beauvarlet in 1760. One, in the Olivier Collection, is signed "Bartholony à Paris" (inv. OA 8346); another, signed "Baillon à Paris," was part of the famous Jubinal de Saint-Albin Collection; a third, unsigned, has been reproduced in the review, *Connaissance des Arts*. A snuffbox in the Rijksmuseum in Amsterdam also presents this Fragonard painting (inv. 17141).

Engravings after Boucher were often used by enamelers; scenes from the series of little cupids particularly inspired them. Among the prettiest offerings of the kind, is the watch in the Olivier Collection signed by Julien Le Roy, representing three cupids playing among flowers on a gray-over-rose camaieu (inv. OA 8374).

Even when they are copies, these paintings have seductive qualities for the art connoisseur. The bright luster of unchangeable

142 Round watch. Painted enamel gold case. Movement signed "Juln Le Roy A Paris." Second quarter of the eighteenth century. D. 47; Th. 25.
Springtime, a work by P.-J. Cazes (1676–1754), engraved around 1720 by Desplaces, is reproduced on the bottom. A circular landscape decorates the bezel.
Paris, Louvre, Olivier Coll., inv. OA 8455

143 *Left.* Round watch. Painted enamel gold case. Movement signed by Delacrose. Paris, third quarter of the eighteenth century. D. 40; Th. 20.
The bottom is decorated after a painting by J.-P. Colson (1733–1803), *The Repose*, dated 1759 (Dijon, Musée des Beaux-Arts).
Center. Round watch. Painted enamel gold case. Movement signed "Baillon A Paris." Mid-eighteenth century. D. 30; Th. 18.
A painting on enamel, reproducing *The Well-Advised Children* by G. de Saint-Aubin (1724–1780), decorates the bottom.
Right. Round watch. Painted enamel case. Movement signed "Ageron A Paris." Third quarter of the eighteenth century. D. 39; Th. 21.
The bottom is painted on enamel, after Boucher's *The Education of Love*.
Paris, Louvre, Olivier Coll., inv. OA 8484, OA 8473, OA 8485

141 *The Education of Love*, engraving by Demarteau l'Aîné after a painting by François Boucher (1703–1770) dated 1761 (present location unknown).
Paris, Bibliothèque Nationale, Cabinet des Estampes

142

143

enamel colors, which renders the freshness of the landscapes so well, the skill with which an enameler could introduce, in a very small space, delicacy of composition and subtle variations in the tonality of a scene are particular aspects of painting on enamel indeed worthy of our admiration.

Still-life paintings sometimes decorate the case bottoms. They usually are bunches of flowers, made with roses, carnations, cornflowers and daffodils, arranged in vases or baskets, placed on a marble pedestal or a table. Sometimes fruits are set next to the bouquets.

Rococo motives on engraved watchcases

Unenameled metal cases were engraved, chased or repoussé with the greatest care. Many were asymmetrically ribbed in the manner of a shell. This very modish decoration was inspired by models that Justin-Aurèle Meissonnier published in his *Livre d'ornements* (Book of ornaments), engraved by Huquier. The artist proposes a very elegant rococo decoration in the engraving of watchcases. On some of the cases, huge waves seem to support ancient structures; on others, volutes of foliage, embedded with flowers, seem to whirl around.[92]

Repoussé work was not as fashionable in France as in England. It is difficult to find French watches with an entirely repoussé case.

Floral decoration uniting engraving with enamel painting appears to have been particularly favored in the mid-eighteenth century. The flowers, ususally roses, were painted in a naturalistic manner on a ground engraved with parallel waved lines. In the same style, a branch of a flowering rosebush decorates the case of a watch by J.-J. Gudin that is now, with its matching chatelaine, in the Olivier Collection (inv. OA 8394). A watch signed by Julien Le Roy, also in the Louvre (inv. OA 2338) is also enameled with roses. A watch of J.-B. Baillon, decorated with a similar rose bouquet, belongs to the Musées Royaux d'Art et d'Histoire in Brussels.[93]

Dial and movement decoration

The rather plain dial is in white enamel; the hours are indicated in Roman numerals and the minutes in Arabic numbers. The watchmaker's signature is sometimes painted in the center.

The cock is no more than two centimeters in diameter; it has retained the extended ears it had in the time of the Regency. Decorated with care and elegance, it is gilt and pierced, engraved with florets and foliage, whose asymmetrical curves and countercurves are in the style of the period. It is sometimes pierced, making a latticework with many floret-linked lozenges.[94]

144 *Music* (La Musique), engraving by Etienne Fessard, dated 1756, after a painting by Carle Van Loo (1705–1765), intended for the Château de Bellevue, now in the Palace of the Legion of Honor, San Francisco.
Paris, Bibliothèque Nationale, Cabinet des Estampes

Watch accessories

Watches are often protected with a second case covered with sharkskin or studded leather. The use of the second cases was less widespread in France than in England.

The fashion of hanging watches from chatelaines appeared toward 1740 and was abandoned only in the nineteenth century.

145 Round watch. Engraved and painted enamel gold case. Movement signed by Baillon. Paris, around 1760. D. 45.5.
The bottom is decorated with the Van Loo painting, *Music.*
Toulouse, Musée Paul Dupuy, inv. 18258

145

146 Model of a watch matched by a chatelaine. Engraving by Jean Mondon, selected from his *Premier livre de pierreries pour la parure des dames* (First book of gems for women's finery). Paris, around 1740.
Paris, Bibliothèque Nationale, Cabinet des Estampes (Le 47 a)

Chatelaines are made of three or four ring-linked plates. From the first plate, ornamented with an escutcheon, hang little chains suspending the watchkey and various trinkets, such as seals and vials. The chatelaine is decorated to match its watchcase.

The years 1760–1790: the neoclassical style

Painting over enamel and engraving continued to be much used in watch decoration. In contrast to the traditionalism of the preceding century, new subjects appeared on the cases; new ways of utilizing enamel permitted the creation of ornaments that became very fashionable.

147 *Left.* Round watch. Case of gold in several colors. Movement signed "Dufalga Genève." Around 1765. D. 42; Th. 20.
Decoration in gold of several colors was often used in watches in the third quarter of the eighteenth century.
Right. Round watch. Case in gold of several colors. Movement signed "Patron A Paris." Last third of the eighteenth century. D. 40; Th. 18.
Two hearts on the altar of love are crowned by a dove. Engravers in the Louis XVI period were often inspired by this type of allegory.
Paris, Louvre, Olivier Coll., inv. OA 8510 and OA 8511

NEW SUBJECTS OF DECORATIONS: THE ANTIQUE MODE, LOVE ALLEGORIES, PLEASURES OF NATURE, PORTRAITS, AND GREUZE'S PAINTINGS

The themes of watchcase decoration are inspired by the contemporary society's taste for antiquity and the pleasures of a rustic life.

Antique scenes in the style of Joseph-Marie Vien were often painted by enamelers. Their clientele seem to have particularly appreciated the *Offerings to Love*. Numerous extant cases indeed represent a couple of young people kneeling in front of Love's altar or a priestess crowning with a flower garland two hearts united on an altar. These scenes, in the bucolic mood so dear to the society of the period, are situated in romantic rural surroundings that harmonize with the lovers' passion.

148 *Left.* Round watch. Enameled gold case. Movement signed "Robin A Paris." Last quarter of the eighteenth century. D. 47; Th. 27.
On the blue enamel bottom, surrounded by brilliants, symmetrical ornaments in the neoclassical style, formed of gold and silver paillons, stand out.
Center. Round watch. Enameled gold case. Movement signed "Revel A Paris." Last quarter of the eighteenth century. D. 38; Th. 14.
The bottom of the case, with a guilloche like the sun, is covered by an opalescent translucent enamel. The bezels are ornamented with enameled cabochons, also imitating opals.
Right. Round watch. Enameled gold case. Movement signed "Lefebvre A Paris." 1786–1787. D. 45; Th. 15.
Suggesting peacock feathers, this chatoyant polychromy was a favorite decoration around 1780.
Paris, Louvre, Olivier Coll., inv. OA 8559, OA 8586, OA 8381

147

148

The love theme plays, at that time, a major part in watch decoration. A watch was certainly a sentimental as well as a useful gift. Love is no longer suggested, as under Louis XV, by amorous encounters, but under a more idealized form, through its symbolic symbols.

Many watches are decorated with an allegory of Love represented by doves, a dog, a quiver, a flowery urn and garlands of roses.

The pleasures of gardening and country life are evoked as well by elements chosen from among everyday objects – a rake, a spade, a basket filled with flowers. To these pleasures are added those of music, hinted at by the presence of a musical instrument – violin, lute or flute.

Watches decorated with these symbolic objects are very often entirely engraved and chased. The four-color gold technique in particular was used in the decoration of that type of case. The watch of a Rouen watchmaker, Duval, in the Olivier Collection, gives a good example of the use of that technique to represent the allegory of Love (inv. OA 8518).

Painting on enamel was used in portraits and in the genres scenes in the manner of Greuze. Usually, the portraits take up a limited space on the case bottom; this evidently explains the banality and the lack of expression common to most of them. The scenes inspired by Greuze's work are skillfully painted; the enameler at times directly copied an engraving of that artist's painting. A watchcase by Baillon, made between 1760 and 1770, is thus decorated with an enamel painting copied after one of Greuze's pieces (New York, Frick collection), engraved by Flipart, entitled *The Wool Gatherer* (Olivier Collection, inv. 8489).

COMPOSITION OF WATCHCASE DECORATION

The subjects are arranged on the cases according to new principles, compared to the preceding period. The engraved motives that surround them belong in the repertory of the Louis XVI style of decorative forms. Like the bezel ornaments, they are generally applied in relief in several colors of gold. The frame at that time occupies an important space in the case bottom, sometimes at the expense of the engraved or painted subject, which is enclosed in a tiny medallion.

A collection in the Cabinet des Estampes in the Bibliothèque Nationale, entitled *Modèles pour les cuvettes de montres* (Models for the cuvettes of watches), presents fashions already popular in 1760, thus helping us recognize their composition.[95] There is no doubt that engravers and enamelers, when not directly copying them, must have found their inspirations in those designs, in which sacrificial scenes, arms or gardening trophies and scattered musical instruments are sometimes accompanied by doves, flowers, a dog or a spring hat. These motives occupy the space of a round or oval medallion in the center of the case bottom. The frame is made of flowers or wreaths of laurel leaves, united by a ribbon topping, and seemingly holding, the medallion. In his plates for goldsmiths published in 1762, Jean-Henri-Prosper Pouget concentrated on the composition of watchcase ensembles and the representation of motifs framing the medallions: draperies, ribbons, flowery branches and laurel wreaths.

SUCCESS OF TRANSLUCENT ENAMEL IN GUILLOCHE

The technique of translucent enamel applied on a machine-turned metal surface successfully renewed the enameled decoration of the cases. It gave them an elegance marked by both simplicity and fantasy.

New forms of decorations were thought out by enamelers; their geometrical compositions were particularly liked by the public. For instance, the bottom of the case, a guilloche of barleycorn or sunrays, was evenly covered with a translucent enamel, on which would stand out an opaque painted enamel rose-window, a scattering of gold stars, stylized flowers made of pearls or diamonds, or motives reminiscent of peacock feathers. This decoration with its chatoyant polychromy, enhanced by golden paillons, seems to have had a certain vogue around 1780; it is found on four watches in the Louvre: one is signed "Lefebvre" (inv. OA 8381), another, "Magnen" (inv. OA 8574), a third, "Mignolet" (inv. OA 8575),[96] a fourth "Lepaute" (inv. OA 9372). The color most in favor was Sèvres blue.

Great care was brought to the edging that encircled the translucent enamel. It is often composed of golden paillons and diamond or half-pearl rosettes – real or white enamel imitations. The rim of the bottom and the bezel holding the glass, were decorated the same as the edging, or enameled with various motives arranged in friezes or in torsades.

DIAL AND MOVEMENT DECORATION

Like those of the preceding period, dials are white enamel. There is a tendency toward simplification. Arabic numerals are no longer saved for the minutes but also often indicate the hours. The traditional numbering of minutes (5, 10, 15, 20, etc.) is replaced by a more discreet numbering of the quarter hours (15, 30, 45, 60) that itself may sometimes be eliminated. Bars and dots compose the minute circle. The hands have a simpler form. The ones in luxury watches are often enriched with small brilliants or small half-pearls.

The cocks, made smaller, are engraved and pierced. They are often formed of four foliage scrollworks, punctuated with florets or geometrical ornaments.

The years 1790–1820: from the Revolution to the Bourbon Restoration

French patrons in this period had very varied tastes. While the "Etruscan" style was supreme, they liked cases decorated with delicate antique scenes. Principles of moderation and severe discipline, vaunted by neoclassical artists, who found in the watchmaking area an illustrious defender, A.-L. Breguet, made them appreciate the simple machine-turned guilloche work, sometimes heightened with transparent enamels. But they also liked fancy watches with their bright polychromy, whose vogue recalls that of form watches in the sixteenth century.

Enamels had a fundamental importance in the decoration of watches at the end of the eighteenth century and the beginning of the nineteenth. The Parisian enamelers and those of Geneva, who furnished French watchmakers with many cases, exploited all the techniques of enamel work.

Enameled decorations

The enamelers at the end of the seventeenth century, often combined painted and translucent enamels with great refinement. Thus, they painted scenes with only a few figures and few accessories standing out on a translucent enamel background. These paintings staged vestals, bacchantes, antique dancing girls and shepherds and shepherdesses and were made in tender colors—pink, sky-blue, light yellow, gray, Nile green—in harmony with their pastoral or elegiac atmosphere. The frame of the scenes was no longer composed of engraved gold ornaments as in the preceding period, but with enameled edgings, underlined with delicate friezes, themselves formed with golden paillons and small opaque enamel motives. The bezels were often decorated with half-pearls. A handsome example of that kind of enameled decoration is offered by André Hessen's watch, made around 1790, now in the Olivier Collection (inv. OA 8593). On the guilloche bottom and covered with a translucent rose enamel, a scene painted in the antique style appears: a bacchante, holding a thyrsus, converses with a maidservant kneeling next to her. A watch with this same scene belongs to the Sandoz collection.

The end of the eighteenth century was marked by a renewal of enamel painting in watch decoration. Watchmakers created a type of flat case with a fairly large diameter whose bottom was occupied by an enamel painting, simply surrounded with a row of pearls. The enamelers often chose to copy the work of a fashionable neoclassical painter. Angelica Kauffmann's *Punished Love* reproduced on the case of a watch of Basile-Charles Le Roy (Olivier Collection, inv. OA 8386).[97]

During the Restoration period, enameled watches were still in great favor. Scenes in the neoclassical taste gave way to flower decorations with a romantic inspiration. Made in champlevé enamels of various colors, the flowers often stood out on a black enamel bottom. Compositions were varied and very elaborate at times. The most common one presented flowers grouped in a bouquet that occupied the center of a medallion. The flowers were arranged in wreaths over a concentrically divided background. Sometimes the background was fan shaped and flowers appeared strewn over it.

Dial and movement decoration

The interest of watches from the revolutionary period sometimes lies in the painted decoration of their dials. Usually, the subjects of that ornamentation were chosen from the imagery of the new ideology. The most often used motives are the Phrygian cap, the fasces standing for union, the olive branch symbolizing peace and the lion representing the people's force. R. Panicali quotes a few mottoes on these dials: "Live Free or Die," "My Life and My Fatherland," "Union, Strength, Freedom and Fatherland."[98]

Watches of that period sometimes have two dials. One marks the time in the traditional way, the other according to the decimal system, as decided by a decree of the Convention on October 5, 1793.[99]

Most watches of the end of the eighteenth century and the first quarter of the nineteenth have a plain white enamel dial. Arabic numerals mark the hours, while Roman ones indicate the minutes. The use of Roman numerals generally only appears on watches having several dials, intended to mark, besides the hour, the date and the age of the moon. In this case, the hours dial is often off-center. The hands in blued steel or in gold, have simple and varied shapes; they can be composed of juxtaposed rings, have arrow shapes or be terminated with fleurons.

Traditional movements, flattened between both plates, retain their gilt cocks, pierced and chased with geometric designs and symmetrically arranged.

149

◁ 149 Round watch. Painted enameled gold case. Movement signed "Hessen A Paris." Around 1790. D. 55; Th. 14.
The harmony of the light tonality of the classically inspired scene is a characteristic of the enamelers' production at the end of the eighteenth century.
Paris, Louvre, Olivier Coll., inv. OA 8593

150 Round watch. Enameled gold case. Around 1830. D. 42; Th. 13.
Champlevé enamels forming a flowery fan ornament.
La Chaux-de-Fonds, Musée International d'Horlogerie, inv. 117.

151 Movement of a decimal watch with a perpetual calendar. Dial signed: "Féron A Paris." 1795. D. 47; Th. 13.
On thermidor 9, year III (July 27, 1795), Andre Féron (1735–c. 1807) let it be known by the Bureau of Consultation on Arts and Crafts that he had invented "a perpetual date indicator for the new calendar." His invention was approved by the Institut de France, on thermidor 6, year VIII (July 25, 1800).
Paris, Musée National des Techniques (CNAM), inv. 1263.

THE BREGUET STYLE

Abraham-Louis Breguet led both a technical and an aesthetic revolution in the watchmaking field. By adopting the caliber invented by J.-B. Lépine, in one stroke he did away with the engraved and chased ornaments of the traditional movement and formed a new aesthetic based on the harmonious relations between the diverse components of the bared wheel works.

Sobriety characterizes the dials of these watches, made in white enamel, silver or gold. Hours are indicated in Arabic numbers on enamel dials, like those of the subscription watches, and in Roman numerals on metal dials, for which the center is generally machine turned. The hands, thin and sharp, often end in a ring. On the enamel or gold dials, they are in blued steel, while on silver dials they are usually gold.

Made in gold or silver, the extra-flat cases created by Breguet have a discreet elegance, conferred to them by their harmonious proportions. The bottom and the band are either smooth or of fine guilloche. Breguet used enamel only for his watches built for the Turkish trade and for some of his *à tact* watches. The pendant is shaped like a ball and lets the suspension ring through. Both components – as well as the hands with their ring endings – subtly recall the purity of the circular shape of the watchcase.

The style conceived by Breguet and notably spread by his hundreds of subscription watches exerted a very large influence, already noticed in the production of his contemporary colleagues, such as Robert Robin, Jean-Antoine Lépine, Basile-Charles Le Roy and Charles Mugnier.

UNUSUAL WATCHES

Appearing in the last quarter of the eighteenth century, unusual watches were very much in fashion during the Empire and the Restoration. Seldom manufactured by French watchmakers, they were in general imported from Switzerland.

Their fastidiously enameled cases borrow their shapes from musical instruments like the lyre, harp or mandolin, from flowers, from fruit and from seashells.

Automaton – often musical – watches, present mannikins striking the time on a bell, a spinstress, a knife-grinder, rope-jumpers, musicians or dancers.[100]

The English watch

The infatuation of English society for rococo under the reign of George II (1727–1760) and later for the neoclassical style under the reign of George III (1760–1820) is particularly well reflected in the decoration of watches.

The English watch kept the pair case, which so clearly distinguished it from its French counterpart. The largest part of the decoration was saved for the outside casing and so it was at times necessary to protect it with a third case.

The English watch and the French watch differed not only in the composition of their cases, but also in their techniques and their decorative styles. Thus, English artists interpreted the rococo style with much more exuberance than the French. Throughout the century they used decorative processes little used on the continent, such as horn painting. They seldom used enamel painting and the technique of four-color gold, which were in a greater vogue in France.

The years 1730–1770: the rococo style

The rococo epoch was a fortunate period for English watch decoration. From around 1730 the outside cases received such a profusion of rococo ornaments that they can be linked to the creations in the Chippendale style. A few decorative forms, only exceptionally used on the continent, were in great favor, repoussé decoration in particular.

152 Pair case round watch. Repoussé gold outside case signed: "Moser fecit." Movement signed "Jn° Ellicott London 3719." Around 1750.
The watch carries two prestigious names: the one of John Ellicott (1706–1772), watchmaker to the king, and the one of a goldsmith attached as a draftsman to George III, George Michael Moser (1705–1783).
London, Victoria and Albert Museum, inv. M 63–1954

153 Repoussé gold round case signed "H. Manly fec." Around 1730.
In the center of a rococo frame, a scene shows Aeneas welcomed by Dido. The circumference is ornamented with eight cartouches, decorated with episodes from the *Aeneid.*
London, Victoria and Albert Museum, inv. 288–1854

154 Round watch. Repoussé case signed by John Gastrell and dated 1762 (hallmark). D. 48.
London, British Museum, inv. CAI 265

155 Pair case round watch. Repoussé gold outside case. Movement signed "Sam Altkins London." 1748–1749. D. 45; Th. 19.
A scene representing Hippomenes and Atalanta decorates the bottom.
Paris, Louvre, Olivier Coll., inv. OA 8493

152

153

154

155

Vogue of the metal repoussé scenes

The process of repoussé appeared rather frequently on metal cases after around 1715. It was used at that time simultaneously with engraving and chasing work. The case band was engraved and pierced with a symmetrical decoration of scrollwork and medallions, in the traditional style of the end of the eighteenth century. Surrounded by symmetrical volutes, the repoussé scene occupied the central part of the bottom.

This decorative mode became more and more important and finally dominated all other forms of ornamentation in the mid-eighteenth century. It was generally applied to gilt metal or gold cases. First executed in low relief, repoussé decoration progressively became so deep that some figures almost appeared freed in high relief from the base. The cases between 1720 and 1730 offer scenes symmetrically framed with arabesques ending as volutes. Around 1740 the framing ornaments, borrowed from the rococo style repertoire, began being increasingly used in asymmetrical compositions. They were inspired by models published by the French engraver J.-A. Meissonnier and the English goldsmith George Michael Moser.

Born in Switzerland in 1706, G. M. Moser came to London in 1720, where he married in 1729–1730. He died in 1783. He specialized in repoussé work, as evident from several cases bearing his name. One, in the Victoria and Albert Museum (inv. M 63–1954), made around 1750, represents Diana the huntress in a rococo frame of curves and countercurves. Another, belonging to the Guildhall Museum (inv. no 157), representing Vertumnus and Pomona, contains a movement by Paul Dupin; it is dated 1739–1740.

The names of other goldsmiths specializing in case repoussé work are known thanks to signatures on certain preserved pieces. Henry Manly was active in the second third of the eighteenth century. The British Museum has a case that he decorated (inv. 86, 5–11, 5). The Victoria and Albert Museum owns two of his works: a case ornamented with a scene showing Antony and Cleopatra (inv. 654–1872) and a case richly decorated with episodes from *The Aeneid* (inv. 288–1854). John Gastrell decorated a watchcase bearing the date 1762 in gold repoussé (British Museum, inv. CAI 265).[101]

The subjects of repoussé metal cases are usually drawn from Greek mythology and Roman history. The helmets and armor covered with decoration and draperies with multiple folds indeed permit the interplay of light on the gold surface raised into multiple facets. So as to increase the effects of shadow and light, the artist multiplies the lines underlining each decorative form; thus, the contour of rococo volutes is enhanced by numerous veins. These lighting effects give an intense movement to the scenes, accented by the curves and countercurves of the frame. According to his skill, the *repousseur* increases the number of planes in the scenes: in the first plane, the foreground, he frees the essential figures; in the two others, he shows the secondary participants or the spectators of the action and finally creates a more or less complex architectural background. These metal repoussé scenes truly look like miniature low reliefs.

English cases decorated with repoussé scenes provide particularly interesting evidence of the application in goldsmith's work of the rococo aesthetic based on asymmetry, supple lines and movement. The repoussé decoration united so well with this aesthetic that it knew its greatest glory in the mid-eighteenth century; after 1770, when the neoclassical style began to take root, it went out of fashion.

Cases decorated with hard or precious stones

During the reign of George II, rich Britons loved watches lavishly decorated with hard or precious stones.

Cases ornamented with hard stones always have a very refined appearance. Blood-red jasper and brown or gray agate veined with delicate tonalities are frequently used.

Diverse types of composition, inspired by rococo, were worked out by lapidaries. The stone finely shaped as a cap can be set in a frame engraved with rococo ornaments and in this way form the case bottom. Sometimes the cap is covered with a *résille* of engraved gold upon which volutes and floral wreaths stand out. In

156 *Top.* Pair case round watch. Exterior case in gold and agate, with diamonds, rubies and emeralds. Movement signed "Opiom London." Mid-eighteenth century. D. 43; Th. 20.
Center. Pair case round watch. Exterior case in gold and blood jasper, with diamonds. Movement signed "Cha Cabrier London." Third quarter of the eighteenth century. D. 41; Th. 23.
Bottom. Round watch. Case in gold and red jasper. Movement signed "J Chancey London." Mid-eighteenth century. D. 49; Th. 30.
The three watches, decorated with hard stones, give a good illustration of English watch production in the mid-eighteenth century.
Paris, Louvre, Olivier Coll., inv. OA 8545, OA 8542, OA 8365

other instances, it is composed with a mosaic made of stone pieces separated by a gold armature. The stone is sometimes sculpted. Thus, the blood-red jasper case of a J. Chancey watch, now in the Olivier collection, is decorated with a low relief representing a musician cupid inside a rococo frame (inv. OA 8365).

In all instances, the decoration of the bezel of the case harmonizes with that of the bottom. The bezel may be formed with a simple circular plate, adjusted between engraved gold armatures, or composed of hard-stone medallions alternating with rococo ornaments.

Colored stones and pearls sometimes heighten the decoration of these cases. They composed either frames around the medallions or ornamental floral brackets, standing out in relief over the hard stones. Several watches thus decorated are signed by the watchmaker John Cox. Those in the Musées Royaux in Brussels and in the Sandoz, Wilsdorf and Olivier Collections have an outside case composed of agate medallions framed with small rubies.

The English, more than the French, appreciated cases luxuriously ornamented with precious stones and pearls used in polychrome compositions, generally floral, reflecting the very lively taste in the eighteenth century for combining variously colored stones. The polychromy was accented by the use of a ground of colored enamel or hard stone.

Enameled cases

In the first two thirds of the eighteenth century in England, enamel painting, applied to watch decoration, did not meet with the universal success it received in France and in Switzerland.

However, enameled small luxury objects in the French style, such as snuffboxes, scent bottles, boxes, bibelots offered in "Remembrance of Friendship" or as a "Mark of My Esteem" were very much in vogue.[102] Under the influence of the fashion, enamel shops developed in Birmingham and Bilston. A manufacture founded in Battersea in 1753 produced painted enamels, often inspired by Watteau, Lancret and Boucher and the engravings of S.-F Ravenet. These paintings were reproduced, by a printing process, on the metal surface of objects covered with white enamel. Sales notices prove that the Battersea establishment numbered among its specialties the making of watch cases. Very few of them remain. The Victoria and Albert Museum has one containing a movement signed "Hughes London" (inv. M 4–1957; reserve).

The famous goldsmith G. M. Moser ended his career as an enamel painter. A watch dated around 1765, now in the Guildhall Museum (inv. no 319), having a movement by Conyers Dunlop, is decorated with an antique scene, painted by Moser, representing *Child Hannibal Swearing Vengeance Against the Romans*, from a work by Hubert François Gravelot (1699–1773).[103] The Louvre also keeps a Moser enamel painting decorating a watch of John Ellicott, dated around 1775 (inv. OA 8582). Executed in brown camaieu, the scene represents two young women, draped antique style, next to a small altar.

Decoration of the interior case

The interior case, containing the movement, may be polished, without decoration, or, if it contains a striking movement, may receive an engraved and pierced decoration in the rococo style.

The center of the bottom is in smooth gold, simply engraved with a rose window, a flowering branch, a monogram or a soaring bird. The band is engraved and pierced with foliage and volutes, the unfurling of which is interrupted by two reserved cartouches, engraved with a grotesque mask, a seashell or a small naive landscape with cottages. The number on the movement is sometimes repeated below the pendant, the head of which is often engraved with a rose window.

Dial and movement decoration

In general, the dials are in white enamel and present their numerals in black enamel. Hours are indicated in Roman numerals, minutes in Arabic numbers marked by fives (5, 10, 15, etc.) A circle of vertical strokes completes the numbering of the minutes. "Beetle and poker" hands are often used.

The cock remains the most decorated part of the movement. It is composed of an openwork disk pierced with foliage, a heel often engraved with a grotesque mask or a shell, and an unpierced foot, chased with acanthus leaves and narrower than in the preceding period. The rest of the plate is sometimes covered with another, pierced and engraved with rococo foliage. Pillars shaped like balusters or small columns are the most common.

The years 1770–1820: the neoclassical style

The neoclassical reaction against the asymmetrical lines and the ornamental profusion of the rococo style asserted itself around 1770. English watch decoration reveals a change in style indeed: the cases richly ornamented with repoussé scenes or too profusely decorated with gems were abandoned for much simpler cases where painted and translucent enamels played an important role.

157 Pair case round watch. Exterior case in gold and agate. Movement signed by James Cox. London, around 1760–1780. D. 39.
Medallions in arborized agate, surrounded with rubies and pearls, adorn the outside case. James Cox had made for himself a specialty of that type of decoration.
Brussels, Musées Royaux d'Art et d'Histoire, inv. 2782

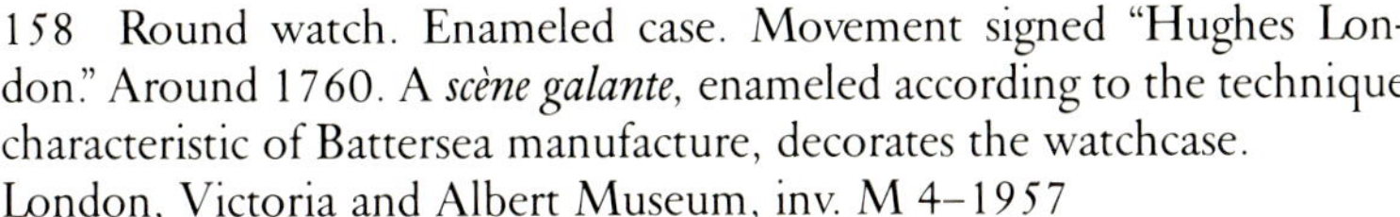

158 Round watch. Enameled case. Movement signed "Hughes London." Around 1760. A *scène galante*, enameled according to the technique characteristic of Battersea manufacture, decorates the watchcase.
London, Victoria and Albert Museum, inv. M 4–1957

159 Round watch. Enameled gold case by G. M. Moser. Movement signed "Conrs Dunlop London 3383." Mid-eighteenth century. D. 50.
A scene showing the child Hannibal taking the oath of vengeance against the Romans, after a work by Gravelot, painted by George Michael Moser on the case.
London, Guildhall Library, The Clock Room, inv. 319

157

158

159

Enameled scenes in the Adam style

Elaborated between 1760 and 1790, the Adam style exerted a strong influence on all decorative arts. Inspired by antiquity, this elegant and delicate style extolled the use of light, symmetrically disposed motives—palmettos, thin garlands, narrow leaves, grotesques—and antique scenes painted in soft and pleasant tones.

When watch decoration receives its inspiration from this style, the cases are ornamented with an oval medallion containing an antique scene, painted in camaieu or in light, soft colors on a translucent enamel base. A disk, also in translucent enamel, underlined by foliage, ovolos, or friezes of Grecian frets surrounds the medallion.

Enamelers sought their models in the work, tinged with sentimentality, of neoclassical painters such as Angelica Kauffmann or Gavin Hamilton. The Victoria and Albert Museum has a watch by James Upjohn whose exterior case is ornamented with a painting after D. Allan's *The Origin of Painting* (1775), kept in the Edinburgh National Gallery, showing a young woman, dressed in antique style, drawing on a wall, by the light of a lamp, the projected profile of her beloved.

160 D. Allan, *The Origin of Painting*, 1775. Oil. Edinburgh, National Gallery

Translucent enamels on a machine-turned guilloche base

As on the Continent, cases decorated with translucent enamels applied on a guilloche base met with a great success in England. They were still used there in the first third of the nineteenth century. Their discreet elegance indeed corresponded to the strong desire for sobriety, asserted as one of the major tenets of neoclassicism.

In order to vary production, many ways of machine-turning the metal were imagined by engravers. Guilloches heightened with subtlety and refinement the transparency of colored enamel, whose gloss aimed to imitate that of precious stones. Half-pearls often decorated the bezel holding the glass and the rim at the bottom.

Tortoise-shell or its imitation decorating the case

A type of case covered with tortoise-shell was received with a favor limited to England during the second half of the eighteenth century and the first quarter of the nineteenth. Most frequently, a thin sheet of transparent horn, painted in imitation of tortoise-shell, covered the case, which was decorated with a scene unvariably framed with holly branches. The horn being fragile, almost all such ornamented cases have cracks. The Victoria and Albert Museum owns several watches of the type: one is decorated with a naive country landscape (inv. M 215–1915, reserve), another with a pastoral (inv. M 279–1919), a third with the device "Hony soit qui mal y pense" (inv. M 212–1915). Watches decorated with painted horn seem to have been a cheap production.

Pendant shape

The elongated pendant with a flat head upon which the oval watch-ring pivoted was used until the end of the eighteenth century. At that time, a new model appeared: the short pendant with a ball-shaped head through which the watch-ring passed. English watches that have been preserved from the first half of the nineteenth century often have a rather heavy "regency" pendant. The oval watch-ring pivots on two screws fixed on the pendant head,

161 Pair case round watch. Enameled gold outside case. Movement signed "Jas Upjohn and Co London." 1778 (hallmark).
The allegory, neoclassical in its inspiration, *The Origin of Painting*, by D. Allan, served as a model for the painted enamel medallion.
London, Victoria and Albert Museum, inv. 1924–1898

which is profiled like an olive; the pendant is slightly convex and has a molded base.

Dial and movement decoration

The white enamel dials are the most common. Simplification and clarity commanded the modifications. Minutes are indicated as discreetly as possible; often the numerals are replaced by a simple circle of dots and vertical bars. The hours are generally marked by long, thin Roman numerals, radially arranged. The dials, in metal guilloche, present the hour indications engraved in circles with a polished surface.

Painted dials often decorate the watches meant for the lower classes. Their scenes, depicted in a naive style, are inspired by farmers', soldiers' or sailors' lives. They occupy the whole surface of the dial or are limited to the center. In the first instance, the hour indications are integrated into the country- or cityscape; a clock in a steeple serves as the hour dial. In the second, hours and minutes are normally marked.

The "beetle and poker" hands were replaced at the end of the eighteenth century by hands known as "spade hands." At first, these hands were very long and thin and only the hour hand had a head. Then they became more robust, and the minute hand swelled up. The end of the hour hand often became heart-shaped. Other types of hands were used by English watchmakers during that period. Serpentine hands, introduced around 1815, were mostly favored by watchmakers from Liverpool. The use of "moon hands" shows the influence of A.-L. Breguet.

Watches for the Eastern market

Watches meant for the Turkish market are, before everything, characterized by a triple case and a dial marked with Turkish numerals. This production includes luxury watches and more ordinary watches, which were exported in large numbers.

The English watchmakers exported watches abundantly ornamented with precious or hard stones to their rich Eastern clientele. The Olivier Collection has two "Turkish watches": one, signed by Peter Dupont of London, is decorated with floral brackets and garnished with pearls, emeralds and rubies over a blue enamel ground (inv. OA 8543); the other, signed "Markwick Markham Perigal London," has an outside case whose agate bottom is covered with an engraved gold *résille* (inv. OA 8546).

Painting on enamel is often used with refinement in the decoration of watches manufactured for the Turkish market at the end of the eighteenth century and during the nineteenth. The case containing the movement and the intermediate case, both ornamented with a landscape or a bouquet of flowers in a festooned frame, are protected by a glass case held by an enameled bezel. A large number of these cases were ordered by English watchmakers from enamel workshops in Geneva.

Watches with more ordinary decoration were also produced for the Turks. They are composed of three gilt brass or silver cases, the last of which is engraved; a fourth case at times ensured a more efficacious protection of the movement against the sand in the desert.[104] This metal fourth case soon became the pretext for an abundantly engraved decoration.

British trade with China was prosperous at the end of the eighteenth century and the beginning of the next. Thus English watchmakers devoted a part of their production to the Chinese market. The watches made by William Ilbery in the first quarter of the nineteenth century are particularly splendid. Their slightest details are ornamented with extreme care. Whether circular or made in diverse shapes, the cases are decorated on the bottom with enamel paintings of excellent quality, copied from fashionable works, either antique in their inspiration or touched with modern sentimentalism. Half-pearls, opaque and translucent enamels and paillons compose the frame of these scenes and ornament the pendant. The movements are as preciously decorated as the cases. Finely engraved foliage scrollwork sometimes stands out from a translucent enamel ground (for instance, red), twisting their shapes all over the surface.

The Swiss watch

The growing importance of enamels marks the evolution of the Swiss watch. During the second half of the eighteenth century and the first half of the next, Genevan enamelers perfected their art to such an extent that they acquired a peerless reputation, particularly in the field of painting on enamel. Watchmaking masters throughout Europe, called on their talents in ordering enamel cases from them for their own watches.

The proliferation of Swiss cases on every world market, containing movements signed not only by Swiss watchmakers but also by their English and French colleagues, gives an international style to watch decoration in this period.

Development of Genevan enamel work

Until around 1770, the French fashions of Louis XV's reign strongly influenced Swiss watch decoration. Enamelers painted scenes in the style of Watteau's *fêtes galantes*, Boucher's pastorals and Greuze's genre scenes. Thus, the Musée Paul Dupuy in Toulouse keeps two watches, signed by Philippe Dufalga of Geneva, ornamented with paintings copied from two Greuze pieces, *Innocence Asleep* and *The Blessing*.[105] In this period, paintings were often situated in the center of a frame of engraved and chased gold rococo motives. The fillet-molded hinge, a characteristic of Louis XV French cases, was adopted by Swiss engravers, as was also the watch-ring pivoting on the pendant head shaped like a flattened stud.

In the last quarter of the eighteenth century, enamel painting tended, as in France, to regain a progressively more important position in watch decoration. The broad engraved gold frames gave way to narrow friezes made of delicate motives such as florets, archings and rows of real or white enamel half-pearls. The preciosity of these borders, often suggesting lace, is accented through the use of foil, either engraved or covered with translucent enamels. The ornaments in the frame often form very elaborate compositions that occupy a large part of the bottom of the case, leaving space to the painter for an often eccentric or fanciful medallion. A character of delicacy and sprightliness is always perceived, in agreement with the subject matter and the treatment of the painted scenes they surround.

Concurring with the cult that the era of Jean-Jacques Rousseau and André Chénier devoted to sensitivity, the enamelers chose sentimental scenes with a bucolic or elegiac inspiration. They used soft and light colors—pink, mauve, gray, beige, pale blue—that extolled the serenity of the atmosphere of the scene and the delicacy of the sentimental exchanges.

To protect their paintings, enamelers perfected around 1780 the "underflux" enamel, also called Geneva enamel. This enamel made a colorless protective glaze that covered the paint.

From the start of the nineteenth century, the role of enamel painting increased. The cases became flatter and larger; their diameters varied from fifty-five to sixty millimeters. The large flat surface of the bottom thus favored the enamel painters' work, exactly as, in the seventeenth century, the large plane faces of the basin-watches did in Blois. In general, the painting occupied the entire

162 *Left*. Round watch. Painted horn case, decorated with a pastoral. *Right*. Round watch. Painted horn case, decorated with a landscape. London, Victoria and Albert Museum, inv. M 279–1919 and M 215–1915

163 Pair case round watch. Enameled gold and engraved outside case, with pearls, emeralds and rubies. Chased and pierced gold case. Movement signed "Petr Dupont London." Mid-eighteenth century. D. 46; Th. 25.
This watch, possessing a dial with Turkish numerals, was intended for Turkey. The bright polychromy that characterizes its case is found on numerous English watches made in the mid-eighteenth century.
Paris, Louvre, Olivier Coll., inv. OA 8543

164 Triple case round watch. Movement signed by Edward Prior. London, first half of the eighteenth century. D. 46.
Because of its triple case, its decoration notably composed of a festooned medallion, its dial with Turkish numerals, the watch is typical of the production intended for the Middle-Eastern market.
Wuppertal, Uhrenmuseum

162△

163△

164▽

165 *Left.* Round watch. Enameled gold case. Movement signed by Jacques Coulin. Geneva, end of the eighteenth century. D. 52.
The case, decorated with an elegiac scene painted in pastel tonalities, enhanced with paillons and pearls, gives a good illustration of taste around 1790.
Center. Round watch. Enameled gold case. Movement signed "Breguet" (the signature is apocryphal). Geneva, end of the eighteenth century. D. 55.
Surrounded by enameled friezes, a *scène galante,* bucolic in its inspiration, decorates the watch and provides a good reflection of the style of the period.

Right. Round watch. Enameled gold case. Movement signed by Patry and Chaudoir. Geneva, end of the eighteenth century. D. 49.
The case is decorated with two *à l'antique* dancers, a decorative subject often used in the neoclassical period.

166 "Chinese" watch. Painted enamel gold case. Movement signed by Bovet. Fleurier, Switzerland, around 1820. D. 58.
This kind of medallion, composed of flowers and painted with a meticulous realism, decorated a very large number of watches intended for the Chinese market.
Wuppertal, Uhrenmuseum

167 △

168 ▽

169 ▽

167 *Left.* Watch with automatons. Anonymous movement. Switzerland, early nineteenth century. D. 50; Th. 22.
The arms of the automaton indicate the hours and the minutes.
Right. "Jacks o' the clock" watch. Anonymous movement. Switzerland, early nineteenth century. D. 52; Th. 19.
Two figures, personifying Abundance and Music, with a sitting child, strike the hours and the quarter hours.
Paris, Musée National des Techniques (CNAM), inv. 21593 and 21968

bottom of the case and was framed simply with one row of half-pearls.

The romantic taste of the time demanded the representation of dramatic scenes drawn from ancient or medieval history. For instance, a painting showing *The Parting of Heloise from Abelard*, from an English engraving, decorates a watch, signed "Piguet et Meylan,"[106] in the Wildsorf Collection.

The style of the enameled paintings in the first third of the nineteenth century is very different from that of the works of the preceding period. The colors are much more pronounced and create stronger contrasts: red, violet, orange, green, etc. The figures, in small numbers, four at the most, take all the space of the painting. They are often shown standing, down to knee level, in calm, almost frozen, attitudes. The decoration is reduced to a few accessories. The undivided attention of the painter concentrates on expressing the psychology of the characters. Watches are no longer painted with charming decorative scenes but are truly miniature paintings executed with a great technical perfection.

The same science of modeling, contrasting light and colors, the same precision in the drawing, the same deep colors are found in the flower paintings that decorated watches until the mid-nineteenth century. Large flowers in full bloom mixed with smaller ones, all bunched close together, form circular bouquets, that occupy almost the whole bottom of the case. Sometimes two doves soar out of these very decorative bouquets.

The oval watch-ring and the pendant, often with a rectangular section, are lavishly garnished with pearls and enamels, harmonizing with the decoration of the frame surrounding the painting and the bezel. The hinge is invisible. The bottom and the rim are opened by secretly pressing tiny studs, half hidden on the band.

168 Watch with automatons. Anonymous movement. Early nineteenth century. D. 67; Th. 24.
On the bottom part of the watch, two cupids strike the hours and the quarter hours on a bell. Above, Moses appears, striking the rock that opens to let out the water with which the Hebrews quench their thirst.
Le Locle, Musée d'Horlogerie, M. and E. Sandoz coll.

169 Anonymous watch movement. Switzerland, around 1840. D. 51.
The center of the dial is enamel painted with a genre scene.
La Chaux-de-Fonds, Musée International d'Horlogerie, inv. 144

Dials and movement decoration

Diverse types of dials followed the white enamel dial in the Louis XV style—this one unvariably marked with Arabic numbers for the minutes and Roman numerals for the hours.

Among the new types, some, with a simple and elegant appearance, indicate the hours in Arabic numbers and no longer carry the minute numerals. Others, more traditional, still present the hours with Roman numerals. The hour circles are sited in various places, often determined by the dial decoration.

Scenes painted either along the circumference or at the central part of the dial are frequent. Their subjects are inspired, in accordance with the revolutionary ideology, by the life of peasants and craftsmen. Dials with automatons had a great success. Instead of presenting an animated scene, watchmakers sometimes preferred to give an automaton the function of indicating the time. In this case, the figure moves his extended arms, used as the watch hands, along two circular arcs marked with the minute and hour numerals.

Among the many varieties of dials that the Swiss watches had at the beginning of the nineteenth century, we note that some still had two hour circles (one in Arabic numbers, the other in Roman numerals), some dials had a pendulum and others had "hour jumpers."

The same variety characterizes the hands. Every shape used by English or French watchmakers was adopted by the Swiss. Hands in the Breguet style met with considerable success at the beginning of the nineteenth century.

The majority of watches signed by Swiss watchmakers have, in the French manner, circular cocks. Ornamented with rococo scrollwork in the second third of the eighteenth century, they are later pierced and engraved with foliage or symmetrically arranged geometric motives.

Two distinctive products

From the start of the nineteenth century, the Swiss were able to gain a virtual monopoly on the production of fantasy watches and watches meant for Oriental markets. One frequently finds the signature of an English or a French watchmaker on watches, which were entirely or partly fabricated in Swiss shops.

◁ 170 Round watch with automatons. Painted enamel gold case. Anonymous movement. Early nineteenth century. D. 56; Th. 21.
Under the painted enamel lid, representing Laban giving his daughter Rachel in marriage to Jacob (?), a scene with musician automatons is hidden.
Le Locle, Musée d'Horlogerie, M. and E. Sandoz.

171 Round watch. Enameled gold case. Movement signed by Louis Roche. Geneva, around 1830.
Champlevé enamels stand out on a black base, forming a decorative composition with feathers, flowers and foliage.
Brussels, Musées Royaux d'Art et d'Histoire, inv. G 857

172

◁ 172 Round watch. Painted enamel gold case. Movement signed by Courvoisier & Co. La Chaux-de-Fonds, around 1820. D. 60; Th. 20.
As evidenced by its Turkish numerals, the watch was intended for the Middle-Eastern trade.
La Chaux-de-Fonds, Musée International d'Horlogerie, inv. 1528

173 Repeater watch movement, with music made by a comb and a notched cylinder. Anonymous. Switzerland, around 1820. D. 56; Th. 12.
Paris, Musée National des Techniques (CNAM), inv. 10652

174 Watch in a perfume-spraying pistol. Anonymous movement. Geneva, early nineteenth century. L. 130; D. of the watch movement 16.
The flower's pistil that surges when the trigger is pulled is pierced with holes to permit the spraying of the perfume.
Le Locle, Musée d'Horlogerie, M. and E. Sandoz.

173

174

175 Form watches. Switzerland, nineteenth century.
Lorgnette, gold, enamel, diamonds. L. 84.
Butterfly, gold, enamel, pearls, diamonds. W. 36.
Mandolin, gold, enamel. L. 64.
Harp, gilt silver. W. 90.
Ladybug, gilt metal, enamel, diamonds. L. 46.
Rockford (Illinois), Time Museum, inv. L 219, B 176, D 170, 144, L 265

Unusual watches

Form watches

Reminiscent of the sixteenth-century fantasy watches, the form watches produced in Switzerland from the end of the eighteenth century also borrowed their shapes from the flora, animal kingdom, fruits and objects of the daily life. Some enjoyed a particular favor, such as the mandolin watches. The forms of the lyre and the harp also inspired manufacturers fairly often. Tulip buds, roses, daisies, apples, pears, floral baskets, urns and seashells made fine cases as well, real jewels that were worn as pendentives or pinned as brooches.

Their precious and refined appearance is most often due to a skillful composition that unites the chatoyant effect of translucent enamel on a guilloche ground with opaque enamel and chased, engraved gold. The chosen colors are lively and form bold contrasts—blue and red, green and red—whose luster is enhanced by reserved gold and finely chased motives. Pearls, diamonds, and colored stones are lavished on the most luxurious watchcases.

The watchmakers' clientele also appreciated, notably during the Empire period, ladies' opera glasses, snuffboxes, poniards or pistols spraying perfumes.

Automatons and musical watches

The inventive and whimsical spirit of Swiss watchmakers revealed itself quite remarkably in the manufacture of tiny luxury objects such as mirrors, scent bottles, opera glasses, snuffboxes, pendentives and watches, enlivened by mechanical figures.

In watches, the automatons are disposed either on the dial face or on a plate hidden under the cuvette of the case. Generally made of colored gold, they stand out on an engraved or enameled decorative base.

The simplest and most frequent scene represents one, two or several figures striking the time on a false bell. The animation of these mannikins is prompted by the repeater, whose ringing is produced, in reality, by hammers striking a bell fixed in the cuvette of the case or by bell-springs placed around the movement.[107]

A good many other animated scenes, competing in ingenuity and originality, were imagined. Some were inspired by the work of craftsmen, such as carpenters, blacksmiths, spinstresses, etc. Among these, the public very much liked the figure of the knife grinder, which often took the form of an arrow-sharpening cupid in the company of other blacksmith cupids. Others took their inspiration from mundane diversions or circus amusements. The action of musical automatons is accompanied by a tune played by a drum, revolving disk or cylinder whose picks cause the blades of a keyboard to vibrate. Placed in a rural setting or a bourgeois home, these musicians—lute, harp, lyre or violin players—often accompany either dancers or a tightrope walker perilously dancing in time to the music. A few animated scenes are borrowed from religious history. One, decorating a particularly splendid watch in the Sandoz collection, represents Moses striking the rock. In the first position, the patriarch gets the rock to open and let the water out. The water, seen quenching the thirst of a few Hebrews, is made of revolving twisted glass. In the second, Moses has the rock close back. The scene stands out in colored gold on an overenameled ground showing the Hebrew crowd.

After around 1800, mechanical musical devices were introduced with a growing success in small-scale objects, often enlivened by automatons. Their invention is attributed to Antoine Fabre (1734–1820), a Genevan watchmaker. He may have constructed his first model in 1796, which played the popular tune "J'ai du bon tabac . . ." Geneva rapidly became a center specializing in the manufacture of these luxurious musical objects. Though most are anonymous, the masters who made these masterpieces are known. Among them, we can name Isaac Piguet, Philippe Meylan and Henri Capts.

Watches for the Eastern trade

From the second half of the eighteenth century, the Swiss watch trade increased with Turkey, China and India.

At the end of that century, Swiss watchmakers had dangerous English competitors on the Turkish market. But as we already mentioned, numerous watches sold to the Turks by the London watchmaker-merchants had their cases decorated in Switzerland. Swiss concerns like Robert & Courvoisier were able to force their products on the whole of the Turkish Empire.

Watches meant for Turkey are easy to recognize. Their white enamel dial has black enamel Turkish numerals. They are made with three and sometimes even four cases. The luxury watches are lavishly enameled. The enamel paintings have particular characteristics responsive to Turkish tastes. Very often, they represent ver-

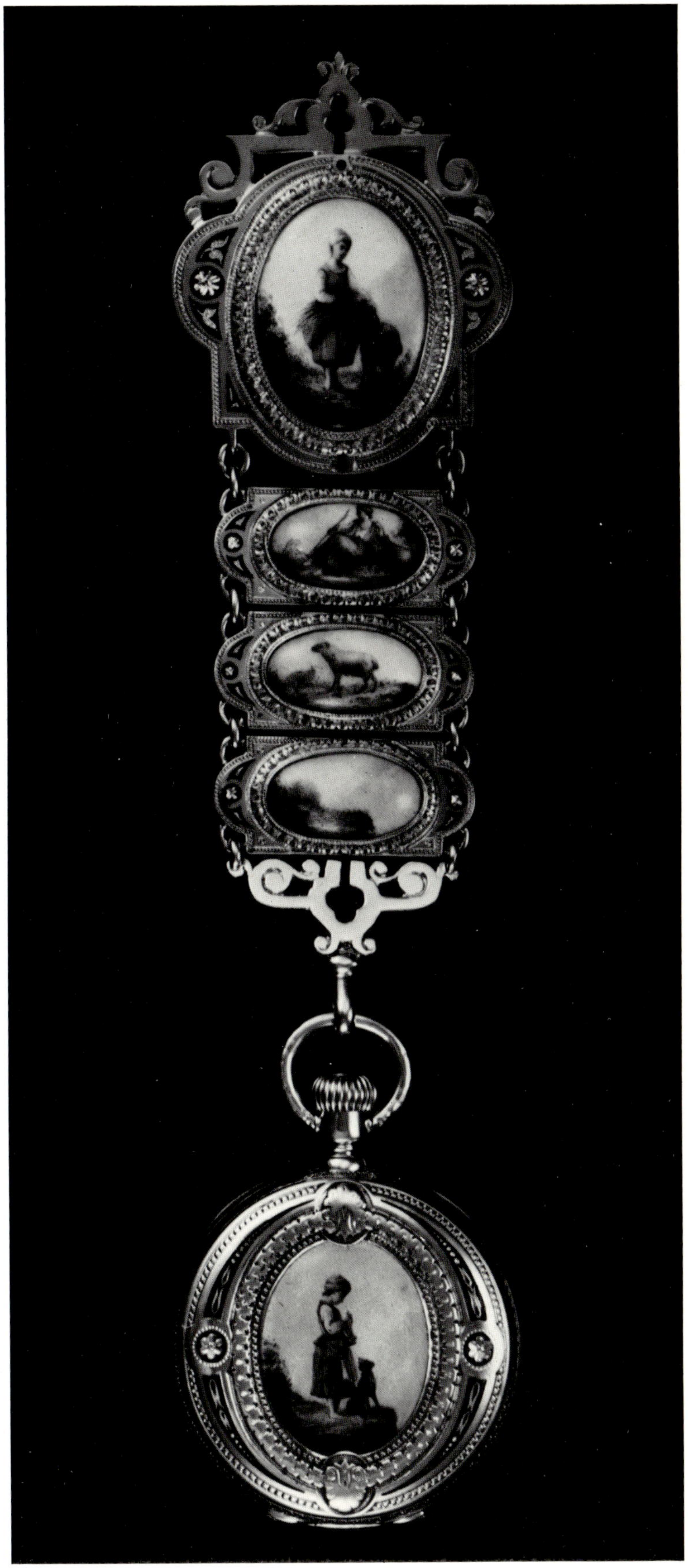

178 Watch with its chatelaine. Engraved and enameled gold case and chatelaine. Movement by Patek Philippe. The piece, bearing the number 32216, was sold on September 11, 1872. It illustrates well the taste of nineteenth-century clientele for the Louis XV style.
Geneva, Patek Philippe

179 Round watch. Engraved gold case, signed "J^s L^s Jacot." Movement signed "F Bovet A Bienne." End of the nineteenth century. D. 57; Th. 20. A scene copied after a Leopold Robert (1794–1835) painting, *The Arrival of the Harvesters in the Pontine Marshes,* decorates the case bottom. Jules-Louis Jacot shows a great talent as an engraver through the precision and the delicacy of his drawing and the expert rendition of light and shades.
La Chaux-de-Fonds, Musée International d'Horlogerie, inv. 171

180 Round watch with key. Gold case, paved with half-pearls. Anonymous movement.
Zurich, Museum der Zeitmessung Beyer

watches decorated with genre and religious scenes, portraits, landscapes, flowers and foliage. In 1895 more than six hundred engravers were active in La Chaux-de-Fonds, about two hundred of them working in Bienne.[111]

One of the best examples of their talents is a watch in the Musée International d'Horlogerie (inv. no. 171), its case engraved by Jules-Louis Jacot, from a famous painting by Léopold Robert (1794–1835), *The Arrival of the Harvesters in the Pontine Marshes*, now in the Louvre. The skillful interplay of shadows and light suggesting perspective, the precision and the flexibility of the contours, the sensitivity of the drawing and the subtle adaptation of the decoration to the shape of the case are as much as the qualities that define the work of the engravers of Geneva and of the canton of Neuchâtel in the nineteenth century.

To this production, often characterized by an abundance of ornamentation, is opposed, during the entire century, to a production born of the style developed by Abraham-Louis Breguet, extolling simple forms, emphasizing decorations that cast off decorative exuberance but did not exclude refinement. The gold case, studded with pearls, of a watch in the Museum der Zeitmessung (Beyer) in Zurich, offers a good example of this style.

In a general way, a decline in the decorative quality of the watches can be noted around the mid-nineteenth century. Through the initiatives of Parisian jewelers and Art Nouveau artists, watches regained, little by little, at the end of the century and the beginning of the next, the elegance that had characterized them. It is curious to see certain decorations embellishing them again; one notes, for instance, the use by Cartier of transparent enamels applied on a radial guilloche ground. Nowadays, watchmakers and jewelers strive to give their watches the style of our times, by using unembellished forms in creating simply engraved or chased bracelets in white or yellow gold and by introducing color, sometimes with much imagination, through the use of precious stones, hard stones and enamels covering and the framing of the dials. Reflecting contemporary fashions with originality and refinement, the manufacture of luxury watches has the same prestige today that it formerly had, increased by the precision brought to the movement by horological discoveries of this century.

Conclusion

The history of small-work horology, both a mechanical and a decorative art, is the history of a technique and an art. It is also the history of a trade.

The successive analysis of these three aspects is indispensable, but it should not deceive the reader in letting him believe that the historical, technical and artistic evolutions are independent. On the contrary, the essential stages observed in each one of them cause reciprocal influences to appear, and the transformations that characterize them often present a close correlation.

From 1500 to 1675 the manufacture of watches appeared mostly oriented toward the production of luxury watches presenting numerous analogies in their decoration to the jewelry of the same period. This production was a response to the demands of a rich clientele, which while considering the watch as an adornment to elegant attire, was at the same time intrigued by the mysteriousness of its mechanism. On the other hand, the watch movement in those days, even though carefully crafted, could not hold claim to much precision: the refined decoration of the container often made up for the defects of the contents. The history of small-work horology went through its most beautiful artistic blossoming in France, in cities like Blois, Lyons and Paris, whose economies and the flourishing of arts and crafts were stimulated by a strong demand for luxury products.

Technical discoveries and historical events precipitated the evolution of watchmaking in the last quarter of the seventeenth century. The emigration of the Huguenots provoked by the revocation of the Edict of Nantes, the centralization of the arts in Paris, the economic crisis and the ossification of guild regulations all caused the decline of French watchmaking to the benefit of the English and Swiss industries. The revolutionary discovery in 1675 of the balance spring gave movements a precision previously impossible to reach and unsettled the industry, all the while making the mediocre situation of the watchmaking art in France apparent vis à vis the technical superiority of its English counterpart. The decoration of watches in that period indeed attests to these upsets: the multiple techniques borrowed from jewelry, goldsmith work and enameling, used during the preceding period in the decoration of watchcases, were abandoned; the greatly varied watchcase forms of the first half of the seventeenth century disappeared to the advantage of the circular shape. From then on, the watch became a functional object. Though often engraved and chased with a great refinement, the cases of the end of the seventeenth and of the first years of the next century have a somewhat austere appearance, partly demanded by the essentially technical preoccupations of the watchmakers of the time.

The artistic and technical renewal of French watchmaking, encouraged by the demands of a luxury-loving clientele and the scientific curiosity of the general public, asserted itself, beginning around 1730. French fashions again spread everywhere. The watchmakers of the French court were famous all over Europe for the technical and artistic quality of their work. This vitality was short-lived. Burgeoning Swiss production became a flood against which few could compete. The French strived with little success to create workshops whose products could satisfy the growing demands of a clientele now accustomed to the low prices of Genevan merchants. No more than France, could England solve the problem of mass production that arose during the second half of the eighteenth century. This weakness of French and English watchmaking benefited the Swiss industry, which, thanks to the commercial dynamism of its *établisseurs* and the extreme division of labor upon which the organization of its production rested, was able to multiply its commercial outlets in Europe and the Middle and Far East. The variety of these outlets explains the diversity of the kinds of watches making up Swiss production: "Turkish watches" for the customers in the empire of the Levant, "Chinese watches" for customers in the Kingdom of Heaven, watches with cases influenced by French decorative styles for their neighbors, form watches, musical watches, automaton watches. This commercial soaring of the Swiss watchmaking to the disadvantage of French and English competition clearly presages its success in the present era.

Appendixes

Notes

Part one

1 C. Fremont, *Origine de l'horloge à poids*, Paris, 1915, pp. 27–28.
2 Jean Lefèvre, Lord of Saint-Rémy. Born in Abbeville toward the end of the fourteenth century, died in 1468. Appointed king-at-arms of the Golden Fleece by Philip the Good. (Michaud, *Biographie Universelle*, t. XXIII.) His portrait is kept in the Musée Royal des Beaux-Arts, Antwerp; cf. *Catalogue descriptif. Maîtres anciens*, 1970, no. 539.
3 The authenticity of the clock has been recognized by experts who examined it (K. Maurice, J. Leopold, K. Pechstein, etc.). See Maurice, 1976, pp. 85–87.
4 Paris, Bibliothèque nationale, 500 Colbert, 127, *Inventaire des joyaulx d'or et d'argent, reliques . . . appartenant à Monseigneur le duc de Bourgogne*, fol. 93 r°.
5 Quoted from V. Gay and H. Stein, *Glossaire archéologique du Moyen Age et de la Renaissance*, Paris, 1928, t. II, p. 33.
6 Planchon, c. 1898, p. 65, Fig. 22.
7 Paris, Archives nationales, KK 51, fol. 97 r°.–v°.
8 *Ibid.*, KK 64, fol. 124 r°.–v°.
9 Arthur de la Borderie, "Inventaire des meubles et bijoux de Marguerite de Bretagne . . . ," *Bulletin de la Société archéologique de Nantes*, Nantes, 1864, t. IV, p. 53.
10 A. de Boislisle, *Inventaire de la comtesse de Montpensier*, Paris, 1880, pp. 18, 21.
11 Quoted from Comte de Laborde, who published the inventory of Charles the Bold in his work on *Les Ducs de Bourgogne*, Paris, 1849. Document no. 3140.
12 Bibliothèque royale, Brussels, Manuscript IV 111, fol. 13 v°. H. Michel is the historian who drew attention to this illumination in his "L'horloge de Sapience et l'histoire de l'horlogerie," *Physis Rivista di storia della Scienza*, II, 1960, pp. 291–298.
13 Quoted in Morpurgo, 1954, pp. 68–69. Source: Archivio Gonzagua, F. II. 8, envelope 2340. The letter was published in part by A. Berenzi, *Gli antichi horologi pubblici et Comino de Pontevico*, Cremona, 1907.
14 J. Leopold, *The Almanus Manuscript*, London, 1971 (the clock of the bishop of Mantua, pp. 57–69).
15 Morpurgo, 1954, p. 72.
16 Boislisle, *Inventaire*, p. 15.
17 *Catalogue des actes de François Ier*, Académie des sciences morales et politiques, Paris, 1887–1907, t. 5, p. 429.
18 Quoted in Franklin, 1888, chap. IV. The writer gives as his source *Bibliothèque de l'Ecole des chartes*, 1865, p. 274.
19 *Catalogue des actes de François Ier*, t. 2, p. 18.
20 E. Grésy, "Inventaire, après décès, des bijoux, tapisseries, tableaux . . . composant la succession de Florimond Robertet, ministre de François Ier au château de Bury," *Mémoires de la Société des Antiquaires de France*, t. XXX, 1868, pp. 25–26.
21 Paris, Archives nationales, Minutier central, XIX, 265.
22 *Ibid.*, LXXXVI, 93.
23 From the second half of the nineteenth century, many German historians peremptorily asserted that the inventor of the watch was Peter Henlein (W. Noeldecken, *Peter Hele der Erfinder der Taschenuhren*, 1891; G. Speckart, *War Peter Hele der Erfinder der Taschenuhren?* 1894). A national legend was built, consecrated by two monuments raised in Henlein's memory, one at Glasshutte (1903), the second in Nuremberg (1905). In 1942 a film, *The Immortal Heart*, was produced for his glory in *Ciné-revue*, no. 2566 (03/18 1942).
24 Maurice, 1976, pp. 87–91. About P. Henlein, see also E. Zinner, *Aus der Fruhzeit der Räderuhr*, Deutsches Museum, Abhandlungen und Berichte, Munich, 1954; J. Abeler, *In Sacher Peter Henlein*, Wuppertal, 1980.
25 Paris, Bibliothèque Nationale, Manuscripts, five hundred Colbert, 128, *Inventaire des vaisselles, joyaux, tapisseries, peintures de Marguerite d'Autriche, dressé en son palais de Malines, le 9 juillet 1523*, fol. 93 v°.
26 Quoted in Morpurgo, 1954, p. 68.
27 *Les livres de Hierome Cardanus, medecin milannois, intitules de la Subtilite & subtiles inventions, ensemble les causes occultes et raison d'icelles*, translated from Latin into French by Richard Leblanc, 1556, p. 324.
28 F.A.B. Ward, "Clocks in the Paintings of Titian," *Antiquarian Horology*, March 1985, pp. 224–230.
29 F.A.B. Ward, "The Earliest Illustration of a Watch," *Antiquarian Horology*, December 1964, p. 278; June 1965, pp. 348–349.
30 Quoted in Hayward, 1956, pp. 3–4.
31 Morpurgo, 1954, p. 50.
32 J. A. Garcia-Diego, *Los relojes y automatas de Juanelo Turriano*, Madrid–Valencia, 1982.
33 Morpurgo, 1954, p. 76.
34 Examples of this production are now in numerous collections, particularly in the British Museum.
35 Baillie, 1951, year 1565.
36 Eva Groiss, "The Augsburg Clockmakers' Craft," in Maurice and Mayr, 1980.
37 Catalogue of the exhibition *The Clockwork Universe*, Munich (April–September, 1980) and Washington (November 1980–February 1981). Cf. Maurice and Mayr, 1980.
38 Gottfried Mraz, "The Role of Clocks in the Imperial Honoraria for the Turks," in Maurice, Mayr, 1980, pp. 43–44.
39 *Ibid.*, p. 44.
40 K. Maurice (1976) gives a study of these centers, "Zentren des Uhrmacherhandwerks bis zur Mitte des 17. Jahrhunderts" (pp. 127–193).

41 Paris, Archives Nationales, Y6[5] ("Livre jaune grand"); Bibliothèque Nationale, Manuscrits français, 21,795, fol. 253 and 254. Works containing partial or complete publications of the 1544 statutes: Lespinasse, *Les Métiers et Corporations de la ville de Paris*, t. III, p. 549; Raillard (1752); Franklin (1888); Beillard (1907).
42 Paris, Archives Nationales, Minutier Central, CXXII, 28.
43 *Ibid.*, XIX, 107.
44 *Ibid.*, XIX, 265.
45 Paris, Archives Nationales, KK 92, fol. 184.
46 Minutier Central, XIX, 146, 82, 91.
47 *Ibid.*, XIX, 172.
48 *Ibid.*, CXXII, 1027.
49 *Ibid.*, XIX, 267.
50 Reproductions in Brusa, 1978, ill. 57–59.
51 Paris, Archives Nationales, Minutier Central, XIX, 265. The rue des Juifs extended between rue du Roi-de-Sicile and rue des Rosiers. The rue des Arcis occupied the part of rue Saint-Martin from rue de la Verrerie to rue de la Vannerie.
52 *Ibid.*, XIX, 171, 160.
53 *Ibid.*, LXXXVI, 93.
54 *Ibid.*, XIX, 173.
55 *Ibid.*, XIX, 269.
56 *Ibid.*, LXXXVI, 93.
57 *Ibid.*, XIX, 172.
58 Raillard, 1752, p. 36.
59 Félibien, *Histoire de Paris*, t. V, p. 361.
60 Quoted from the manuscript in the Bibliothèque Nationale (cf. note 41).
61 Develle, 1917, pp. 214–224.
62 *Ibid.*, p. 20.
63 *Ibid.*, pp. 263, 298.
64 Cf. the third part of this work.
65 Develle, 1917.
66 *Ibid.*, and Cardinal, 1983, p. 12.
67 Manuscript kept in the Prevote register, Archives Départementales du Loir-et-Cher (series B), published by P. Bourgeois, "Métiers de Blois," *Mémoires de la Société des Sciences et des Lettres du Loir-et-Cher*, t. I.
68 Vial and Cote, 1927, p. 11.
69 *Ibid.*, p. 9.
70 Two clocks with complications signed by Naze are still extant: Paris, Musée du Petit-Palais, Dutuit Coll. no. 1461; Ecouen, Musée National de la Renaissance (reprod. in Cardinal, 1980, p. 8).
71 See in Vial and Cote (1927) the notices on these watchmakers.
72 A. de Charmasse, "L'horlogerie et une famille d'horlogers à Autun et à Genève aux XVIe et XVIIe siècles," *Mémoires de la Société Eduenne*, t. XVI, Autun, 1888.
73 *Ibid.*, p. 179.
74 About the consequences of the Huguenot emigration on the Genevese economy, we consulted Fournier-Marcigny, *Genève au XVIe siècle*, 1942; A. M. Puiz, *Recherches sur le commerce de Genève au XVIe siècle*, 1964. The study published in 1916 by A. Babel, *Les Métiers dans l'ancienne Genève – Histoire corporative de l'horlogerie*, is essential.
75 Babel, 1916, p. 38.
76 Charmase, *op cit.*, pp. 198–202.
77 Tardy, 1972.
78 Paris, Bibliothèque Nationale, Manuscrits français, 21,795, fol. 258–260.
79 Raillard, 1752, pp. 13, 14, 33.
80 Fichier Brateau.
81 Paris, Bibliothèque Nationale, Manuscrits français, 21,795, fol. 269–272.
82 Develle, 1917.
83 *Ibid.*
84 Paris, Bibliothèque Nationale, Mélanges Colbert, 75, fol. 6 and 7.
85 *Inventaire de tous les meubles du cardinal Mazarin dressé en 1653*, London, 1861, p. 62.
86 Paris, Archives Nationales, KK 191.
87 Develle, 1917, p. 32.
88 *Ibid.*, p. 35.
89 Cf. Vial and Cote, 1927.
90 *Ibid.*, p. 22.
91 *Ibid.*; see notices on these watchmakers in Vial and Cote.
92 Cardinal, 1983, p. 16.
93 Cardinal, 1984, notice 5.
94 Tardy, 1972, repro. p. 25.
95 Statutes published in Babel, 1916, pp. 548–551.
96 *Ibid.*, p. 56.
97 Chapuis and Jaquet (1945, p. 24) consider this watch the oldest known to carry a Genevan watchmaker's signature.
98 Vial and Cote, 1927. Fichier Brateau.
99 P. J. Shears, "Huguenot Connections with the Clockmaking Trade in England," *Huguenot Society Proceedings*, 1960.
100 Hayward, 1956, p. 5.
101 Cf. the third part of the present work.
102 Atkins, 1881, pp. 2–3.
103 Develle, 1917.
104 Atkins, 1881, pp. 7–19.
105 *Ibid.*, pp. 22–49.
106 Cipolla, 1967, p. 132.
107 *Ibid.*
108 Scoville, *The Persecution of Huguenots and French Economic Development 1680–1720*, Berkeley, Los Angeles, 1960.
109 Vial and Cote, 1927, p. 20.
110 Scoville, *op cit.*, p. 125.
111 "Horlogers protestants établis à Sedan." *Société de l'histoire du protestantisme français*, 1890, t. 39, pp. 562–563. Acts relating to J. and D. de la Feuille are kept in Amsterdam, in the Archiefdienst der Gemeente.
112 Morpurgo, 1970. Fichier Brateau.
113 Scoville, *op. cit.*, p. 123.
114 *Ibid.*, p. 322.
115 Research in the archives of the London Goldsmiths' Guild for the period 1680–1720 revealed that there were at least 146 goldsmiths, jewelers and lapidaries among the French refugees (Scoville, *op. cit.*, p. 330). It would be interesting to do the same research in the Clockmakers' Company archives.
116 The list, indicated in the next paragraphs, of French watchmakers who emigrated to England, was done after consultation of the Fichier Brateau and the Baillie Dictionary (1929).
117 Review article quoted in note 99, p. 18.
118 Savary des Bruslons, 1742, t. II, art. "Montre."
119 Scoville, *op. cit.*, p. 125.
120 Fichier Brateau.
121 Paris, Bibliothèque Nationale, Manuscrits français, 21,795, *Avis aux horlogers*.
122 *Mercure de France*, January 1719, "Relation abrégée de l'établissement de la nouvelle fabrique d'horlogerie à Versailles."
123 Savary des Bruslons, 1742, t. II, art. "Montre."
124 Paris, Archives Nationales, X[1A] 9161 and 9162, fol. 113. The watches

have the inv. nos. 1239 to 1242.
125 Many works contain the biographies of the great English watchmakers. Among those, we found Chamberlain's (1941) particularly useful, as well as the 1973 edition of *Britten's Old Clocks and Watches* . . .
126 Babel, 1915, pp. 81–85.
127 *Ibid.*, chap. V, "Les femmes dans la Fabrique."
128 *Ibid.*, p. 391.
129 *Ibid.*, p. 96.
130 *Ibid.*, p. 101.
131 Chapuis and Jaquet, 1945, p. 29.
132 J.-J. Rousseau, *Confessions*, t. I, p. 7, Livre de Poche ed., 1972.
133 J. Starobinsky, *L'Invention de la liberté*, Paris, 1964.
134 R. Mandrou, *La France aux XVIIe et XVIIIe siècles*, Paris, 1967, p. 78.
135 Casanova, *Mémoires*.
136 Paris, Bibliothèque Nationale, Manuscrits français, 14,112–14,113.
137 Paris, Archives Nationales, 0^{1} 2984.
138 *Op. cit.*, note 136, p. 336.
139 Paris, Archives Nationales, 0^{1} 3254.
140 L. Latzarus, *Beaumarchais*, Paris, 1930.
141 *Livre-Journal de Lazare Duvaux, marchand-bijoutier ordinaire du Roi (1748–1758) (. . .)*, Paris, Société des Bibliophiles français.
142 It meant a second case. The use of paired cases was general in the English production. See the third part of the present work.
143 C. Cardinal, pp. 271–273, in *Ferdinand Berthoud*, La Chaux-de-Fonds, 1984.
144 Verlet, 1970, p. 86.
145 Christian Bualez, "Notes sur quelques meubles et objets d'arts des appartements de Louis XVI et de Marie-Antoinette," *Revue du Louvre*, no. 5/6, 1978, p. 371.
146 Paris, Bibliothèque Nationale, Manuscrits français, 21795, fol. 287 to 289.
147 *Ibid.*, fol. 300 ff.
148 J. Le Roy, "Mémoire contenant les moyens d'augmenter le commerce et la perfection des ouvrages d'horlogerie," in *Manuscrits de Julien et Pierre Le Roy*," Musée National des Techniques (CNAM), Archives 4° 126.
149 Raillard, 1752.
150 Savary des Bruslons, 1742, t. II, art. "Horloger."
151 Paris, Archives Nationales, 0^{1} 1672, nos. 133, 135.
152 Diderot et d'Alembert, 1751–1780, t. VIII, pp. 306–307.
153 J. D. Augarde, "L'Atelier de Ferdinand Berthoud," in *Ferdinand Berthoud*, La Chaux-de-Fonds, 1984.
154 *Ibid.*
155 Gélis, 1950. Gélis refers to an article by Charles Guillou, "Ecole et manufacture royale d'horlogerie de Bourg-en-Bresse, 1764–1777," *Feuilleton du Courrier de l'Ain*, Oct. 11–Nov. 20, 1904.
156 *Ibid.*
157 Beillard, 1895, chap. XII.
158 *Rapport et projet de décret sur la manufacture d'horlogerie de Besançon*, by Boissy d'Anglas, Nivôse, year III (1795), Imprimerie Nationale.
159 Tardy, 1967, p. 170.
160 Barbey, 1915, p. 5.
161 Tardy, 1967, p. 171.
162 Diderot and d'Alembert, 1751–1780, t. VIII, p. 310.
163 Tardy, 1972.
164 Diderot and d'Alembert, 1751–1780, t. VIII, p. 310.
165 Claude Frégnac, *Les Bijoux*, Paris, Hachette, Plaisir des Images coll., 1966, reprod. p. 86.
166 Tardy, 1972, p. 252.
167 J. J. Fiechter, *Un Diplomate américain sous la Terreur*, Paris, 1983, pp. 28 and 37.
168 Le Roy, *op. cit.* (note 148).
169 *Ibid.*
170 Paris, Bibliothèque Nationale, Manuscrits français, 21,795, fol. 300 ff.
171 Savary des Bruslons, 1761, art. "Horlogerie."
172 *Ibid.*, "Publication des statuts de la Corporation genevoise," pp. 338–341.
173 Babel, 1916, pp. 81–85.
174 Chapuis, Jaquet, 1945, p. 95.
175 *Ibid.*, p. 52.
176 *Ibid.*, p. 160.
177 *Ibid.*, p. 54.
178 Savary des Bruslons, 1761, pp. 338–341.
179 Chapuis, Jaquet, 1945, p. 134.
180 The essential work on the subject is A. Chapuis, *La Montre Chinoise*, 1919. The author devotes a chapter to Charles de Constant (below), entitled "Un commerçant suisse en Chine à la fin du XVIIIe siècle: Charles de Constant" (chap. IV).
181 Chapuis, 1919, p. 95.
182 *Ibid.*, p. 133.
183 Tardy, 1967, p. 168.
184 Babel, 1916, chap. VI, "Grands ateliers et machines-outils."
185 Atkins, 1881, p. 129.
186 *Ibid.*, p. 97.
187 Cf. the second part of the present work.
188 Chamberlain, 1941, and Baillie, 1973.
189 Archives Nationales, 0^{1} 1919, fol. 261.
190 Le Roy, *op. cit.* (note 148).
191 Kurtz, 1975, p. 73.
192 Chapuis, 1919, p. 54.
193 *Exposition universelle de 1889, à Paris. Rapport du jury international*, A. Picard, director, Paris, 1891. "Classe 26, Horlogerie," by P. Garnier, p. 693.
194 *Exposition universelle de 1889, Congrès international de chronométrie, Comptes-rendus des travaux*, Paris, 1891. "Notes sur l'horlogerie," by P. Garnier, pp. 39–40.
195 Garnier, *op. cit.* (note 193), pp. 710–711.
196 *Ibid.*
197 In 1876, after the Philadelphia World Exposition, the Swiss delegates published a report to alert their countrymen to the danger of American competition.
198 E. O. Lami, *Dictionnaire encyclopédique et biographique de l'industrie et des arts industriels*," Paris, 1881, p. 690.
199 *Etudes ou rapports sur l'Exposition de 1878*, E. Lacroix, Paris. La Suisse, pp. 514 ff.
200 *Ibid.*
201 *Ibid.*
202 *Ibid.*

Part two

1 Develle, 1917, p. 171.
2 Paris, Archives Nationales, KK 147 (n.p.).

105 Gélis, 1950, pl. LI.
106 Chapuis, 1944, pl. 22.
107 The best study on automatons remains the one by Chapuis and Gélis, *Le Monde des automates*, 1928. Chap. 17: "Montres et tabatières."
108 See on this subject the basic study by Chapuis, 1919.
109 Chapuis, 1919, p. 199.
110 Cardinal, 1983, *La décoration des ouvrages d'horlogerie et le goût artistique du XIXe siècle,*" pp. 102–117.
111 *Ibid.*, p. 98.

Bibliography

Books and articles

Abeler, J., *Meister der Uhrmacherkunst*, n. p., 1977.

Atkins, S. E., Overall, W. H. *Some Account of the Worshipful Company of Clockmakers of the City of London*, London, 1881.

Babel, A., *Les Métiers dans l'ancienne Genève. Histoire corporative de l'horlogerie, de l'orfèvrerie et des industries annexes*, Genève, 1916.

Babel, A., *La Fabrique genevoise*, Neuchâtel, Paris, 1938.

Baillie, G. H., *Watchmakers and Clockmakers of the World*, London, 1929, 2d ed., 1947, 3d ed., 1951. Reprinted 1976.

Baillie, G. H., *Watches*, London, 1929. Reprint, 1979.

Baillie, G. H., *Clocks and Watches: An Historical Bibliography*. London, 1951, 3d ed. reprint, 1978.

Baillie, G. H., Clutton C., Ilbert C. A., *Britten's Old Clocks and Watches and Their Makers. An Historical and Descriptive Account of the Different Styles of Clocks and Watches of the Past in England and Abroad, Containing a List of Nearly Fourteen Thousand Makers*, 7th ed., London, 1956, 8th reviewed and augmented edition, London 1973.

Bailly, R., *Les Coqs de montres anciennes*, Clamecy, 1964.

Barbey, F., "La Fabrique d'horlogerie genevoise," Berne, 1915 (excerpt from the *Indicateur d'histoire suisse*, 1915, pp. 175–183).

Bassermann-Jordan, E. von, *Montres, horloges et pendules*, translated from the German from the 4th ed. reviewed and corrected by H. von Bertele, Paris, 1964.

Beillard, A., *Recherches sur l'horlogerie, ses inventions et ses célébrités*, Paris, 1895.

Beillard, A., *La Montre depuis son origine jusqu'à nos jours*, Paris, 1907.

Belmont, H. L., *L'Echappement à cylindre (1720–1950)*, Besançon, 1984.

Berner, G. A., *Dictionnaire professionnel de l'horlogerie*, La Chaux-de-Fonds, 1961.

Berthoud, F., *Essai sur l'horlogerie*, Paris, 1763, 2 vol. Reprinted 1978.

Berthoud, F., *Histoire de la mesure du temps*, Paris, 1802, 2 vol. Reprinted, 1976.

Britten, F. J., *Old clocks and watches and their makers (. . .) in England and abroad, to which is added a list of ten thousand makers*, London, 1904. Consulted ed.: 5th ed, 1922.

Brusa, G., *L'Arte dell'orologeria in Europa*, n. p., 1978.

Bruton, E., *Clocks and Watches, 1400–1900*, London, 1967.

Bruton, E., *Histoire des horloges, montres et pendules*, Paris, 1980.

Camerer Cuss, T. P., *The Story of Watches*, London, 1952.

Camerer Cuss, T. P., *The Camerer Cuss Book of Antique Watches*, London, 1976.

Cardinal, C., *Les Montres et horloges*, Rennes, 1980.

Cardinal, C., *L'Horlogerie dans l'histoire, les arts et les sciences, Chefs d'oeuvre du Musée international d'Horlogerie de La Chaux-de-Fonds*, Lausanne, 1983.

Cardinal, C., "Chefs d'oeuvre du Musée international d'horlogerie," in *L'Estampille*, July 1984, pp. 32–41.

Chamberlain, P. M., *It's About Time*, New York, 1941. Reprinted London 1978.

Chapiro, A., "Les Oignons Louis XIV," in *Revue de l'ANCAHA*, March 1976.

Chapiro, A., "Les Oignons Louis XIV. Les Oignons à complications," in *Revue de l'ANCAHA*, September 1977.

Chapiro, A., "Les Oignons Louis XIV. Constructions particulières," in *Revue de l'ANCAHA*, October 1978.

Chapuis, A., *La Montre chinoise*, Neuchâtel, 1919.

Chapuis, A., *Le Grand Frédéric et ses horlogers*, Lausanne, 1938.

Chapuis, A., *A travers les collections d'horlogerie, gens et choses*, Neuchâtel, 1942.

Chapuis, A., "Heures féminines," in *Album du Figaro*, Winter 1947–1948.

Chapuis, A., *De Horologiis in arte. L'Horloge et la montre à travers les âges d'après les documents du temps*, Lausanne, 1954.

Chapuis, A., Gélis, E., *Le Monde des automates*, Paris, 1928, 2 vol. (chap. XVII: Montres et tabatières).

Chapuis, A., Jaquet, E., *Histoire et technique de la montre suisse*, Olten, 1945.

Cipolla, C. M., *Clocks and Culture, 1300–1700*, New York, 1967.

Clouzot, H., "Les Frères Huaud, miniaturistes et peintres sur émail," in *Revue de l'Art ancien et moderne*, October 1907, pp. 296–306.

Clouzot, H., "Documents inédits concernant Jean Toutin et les premiers peintres sur émail", in *Bulletin de la Société de l'Histoire de l'Art français*, 1908, pp. 198–199.

Clouzot, H., "Les Maîtres de Petitot, les Toutin, orfèvres, graveurs et peintres sur émail", in *Revue de l'Art ancien et moderne*, December 1908, pp. 456–466, and January 1909, pp. 39–48.

Clouzot, H., "Les Maîtres de Petitot, les émaillistes de l'Ecole de Blois," in *Revue de l'Art ancien et moderne*, August 1909, pp. 101–116.

Clouzot, H., "Les Maîtres horlogers de Blois," in *Revue de l'Art ancien et moderne*, February 1913, pp. 115–124.

Clouzot, H., "Un Musée de la montre," in *La Renaissance de l'art français*, August 1921, pp. 415–422.

Clouzot, H., "Les Maîtres de la miniature sur émail," in *Gazette des Beaux-Arts*, July–August 1923, pp. 53 ff.

Clouzot, H., *Dictionnaire des miniaturistes en émail*, Paris, 1924.

Clouzot, H., *Les Grandes époques de la miniature sur émail*, 1925 (typewritten text, Bibliothèque Forney, Paris).

Clouzot, H., *La Miniature sur émail*, Paris, n.d., [1928].

Clouzot, H., "L'Heure d'autrefois" in *Plaisir de France*, March 1937.

Clouzot, H., Gélis, E., "Le Décor de la montre du XVIe au XIXe siècle à l'exposition du Musée Galliéra," in *Gazette des Beaux-Arts*, 2d semester 1921, pp. 101–111.

Clutton, C., Daniels, G., *Watches*, London, 1965. 2d ed., 1971. 3d ed., 1978.

Conservatoire national des Arts et Méitiers. Conférences prononcées a l'exposition "Les Chefs d'oeuvres de l'horlogerie," Paris, 1949.

Cote, C., *Montres et horloges*, Lyon, 1919.

Crespe, F., *Essai sur les montres à répétition*, Geneva, 1804.

Cumhaill, P. W., *Investing in Clocks and Watches*, London, 1967.

Daniels, G., *English and American Watches*, London, 1967.

Daniels, G. *The Art of Breguet*, London, 1975.

Defossez, L., *Les Savants du XVIIe siècle et la mesure du temps*, Lausanne, 1946.

Demoriane, H., "La Montre française, un inventaire complet. I. De 1500 à 1675," in *Connaissance des Arts*, no. 172 (June 1966), pp. 70–77.

Demoriane, H., "La Montre française, un inventaire complet. II. De 1675 à 1795," in *Connaissance des Arts*, no. 184 (June 1967), pp. 74–82.

Demoriane, H., "La Montre française, un inventaire complet. III. De 1795 à nos jours," in *Connaissance des Arts*, no. 201 (November 1968), pp. 96–103.

Develle, E., *Peintres en émail de Blois et de Châteaudun au XVIIe siècle*, Blois, 1894. Reprinted 1978.

Develle, E., *Les Horlogers blésois aux XVIe et XVIIe siècles*, Blois, 1913. 2d completed ed., 1917. Reprinted 1978.

Diderot, D., d'Alembert, J., *Encyclopédie or Dictionnaire raisonné des sciences, des arts et des métiers*, Paris, 1751–1780, t. V, article "Emaillerie"; t. VIII, article "Horlogerie."

Dubois, P., *Histoire et traité de l'horlogerie ancienne et moderne (. . .)*, Paris 1850.

Ferrand, J.-P. *L'Art du feu ou de peindre sur émail dans lequel on découvre les plus beaux secrets de cette science (. . .)*, Paris, 1721.

Franklin, A., *La Vie privée d'autrefois*, t. IV, *La Mesure du temps*, Paris, 1888.

Gallon, M., *Recueil de machines et inventions approuvées par l'Académie royale des Sciences*, t. I to IV (1666–1734), Paris, 1735; t. VII (1735–1754), Paris, 1777.

Good, R., *Watches in Colour*, 1978.

Gould, R. T. *The Marine Chronometer, Its history and Development*, London, 1923.

Grandjean, S., "L'Horloge bijou," in *L'Amour de l'art*, 2d trimester 1948.

Gros, C., *Echappements d'horloges et de montres*, Paris, 1913, 2d ed. 1922. Reprinted 1980.

Guye, S., Michel, H., *Mesures du temps et de l'espace, horloges, montres et instruments anciens*, Paris, 1970.

Hayward, J. F., "Watch and Clockmakers Design Books," in *Antiquarian Horology*, March 1955.

Jagger, C., *L'Horlogerie*, Paris, 1973.

Jagger, C., *Histoire illustrée des montres et des horloges*, Pully, 1977.

Jaquet, E., *Horlogers genevois du XVIIe siècle*, Geneva, 1938.

Kurz, O., *European Clocks and Watches in the Near East*, London, 1975.

Lallier, R., "Montres célèbres et anciennes," in *L'Amour de l'art*, 2d trimester 1948.

Lepaute, J.-A., *Traité d'horlogerie*, Paris, 1755.

Lightbown, R., "Les Origines de la peinture en émail sur or: un traité inconnu et des faits nouveaux," in *Revue de l'art*, no. 5 (1969).

Loomes, B., *Watchmakers and clockmakers of the world*, London, 1976.

Maurice, K., *Die deutsche Räderuhr*, Munich, 1976.

Maurice, K., Mayr, O., (. . .), *The Clockwork Universe, German Clocks and Automata, 1550–1650*, New York, 1980 (catalogue of the 1980 Munich exhibition).

Meis, R., *Les Montres de poche*, Fribourg, 1980.

Moinet, M. L., *Nouveau traité général d'horlogerie (. . .)*, Paris, 1848, 2 vol.

Montamy (d'Arclais de), *Traité des couleurs pour la peinture en émail et sur la porcelaine précédé de l'Art de peindre sur émail (. . .)*, Paris, 1765.

Morpurgo, E., *L'Origine dell'orologio tascabile*, Rome, 1954.

Morpurgo, E., *Montres précieuses du XVIe au XIXe siècle, avec une esquisse historique de l'horlogerie italienne*, n.d., n.p. [c. 1966].

Morpurgo, E., *Nederlandse klokhen en horlogemachers vanaf 1300*, Amsterdam, 1970.

Panicali, O., *Cadrans de la Révolution, 1789–1800*, Lausanne, 1970.

Patrizzi, O., Sturm, F.-X., *Montres de fantaisie*, Geneva, 1979.

Planchon, M., *L'Horloge, son histoire rétrospective, pittoresque, artistique*, Paris, n.d. [c. 1898].

Planchon, M., *Musée rétrospectif de la classe 96, Horlogerie, à l'Exposition universelle internationale de 1900 à Paris*, 1900.

Raillard, C., *Extraits des principaux articles des statuts des maîtres horlogers de la ville et des faubourgs de Paris*, Paris, 1752.

Reverchon, L., *Petite histoire de l'horlogerie*, Besançon, n.d. [1935]

Salomons, D., *Breguet*, London, 1923.

Saulnier, C., *Traité des échappements et des engrenages*, Paris, 1855.

Savary des Bruslons, J., *Dictionnaire universel de Commerce, d'Histoire naturelle et des Arts et Métiers*, Geneva, 1742, 2d ed. Copenhague, 1761 (articles "Horloger," "Horlogerie," "Montre").

Schneeberger, P. F., *Les peintres sur émail genevois au XVIIe et au XVIIIe siècles*, Geneva, 1958.

Simoni, A., *Orologi italiani dal Cinquecento all'Ottocento*, Milan 1965.

Smith, A., *Clocks and Watches*, London, 1975.

Spierdijk, C., *Horloges an Horlogemakers*, Amsterdam, 1973.

Tardy, *La Montre*, 1st fascicle, *Les Echappements à recul*, 2d fascicle, *Les Echappements à demi-recul*, 3d fascicle, *Les Echappements libres à ancre*, Paris, 1950–1953.

Tardy, *Les Coqs de montres*, Paris, n.d., (1954).

Tardy, "La Manufacture d'horlogerie automatique de Versailles," in *La France horlogère*, September 1967.

Tardy, *Dictionnaire des horlogers français*, Paris, 1972.

Thiout, A., *Traité de l'horlogerie méchanique et pratique, Paris, 1741*. 2 vol.

Verlet, P. *La Mesure du temps*, Paris, 1970.

Vial, E., Cote, C., *Les Horlogers lyonnais de 1550 à 1650*, Lyons, 1927.

Museum catalogues

Basle, Historisches Museum: S. C., L. Nathan-Rupp, *Die Kutschenuhren*, Basle, 1983.

Besançon, Musée des beaux-arts et d'archéologie: P. Mesnage, *Musées de Besançon. Collections d'horlogerie*, Besançon, 1955.
Bruxelles, Musées royaux d'art et d'histoire: A.-M. Berryer, L. Dresse de Lébioles, *La Mesure du temps à travers les âges aux Musées royaux d'Art et d'Histoire*, Bruxelles, 1974.
Geneva, Musée de l'horlogerie: *Musée de l'Horlogerie*, Geneva, 1976.
Jerusalem, L. A. Mayer Memorial for Islamic Art: G. Daniels, O. Markarian, *Watches and Clocks in the Sir David Salomons Collection*, London, New York, 1980.
La Chaux-de-Fonds, Musée international d'horlogerie: *Collections du Musée international d'horlogerie, La Chaux-de-Fonds, Suisse*, La Chaux-de-Fonds, 1974.
Le Locle, Musée d'horlogerie: *Collection d'horlogerie de la ville du Locle, Château des Monts*, Le Locle, n.d. [c. 1975]
London, British Museum: H. Tait, *Clocks and Watches, British Museum*, London, 1983.
London, Guildhall Museum: C. Clutton, G. Daniels. *Clocks and Watches in the Collections of the Worshipful Company of Clockmakers*, London, 1983.
London, Science Museum: F. A. B. Ward, *Science Museum. Descriptive Catalogue of the Collection Illustrating Time Measurement*, London, 1966.
London, Victoria and Albert Museum: J. F. Hayward, *English Watches, Victoria and Albert Museum*, London, 1956. 2d ed. 1969.
Madrid, Museo Grassi: L. Monreal y Tejada, *Reloges antiquos (1500–1850)*, Madrid, 1955.
Milan, Museo Poldi Pezzoli: G. Brusa, *Gli Orologi, Museo Poldi Pezzoli*, Milan 1974.
Oxford, University of Oxford Museum of the History of Science: F. Maddison, A.-J. Turner, *University of Oxford Museum of the History of Science. Catalogue 2: Watches*, Oxford, 1973.
Paris, Musée du Louvre: C. Cardinal, *Les Montres du Musée du Louvre. Tome I, La collection Olivier*, Paris, 1984.
Paris, Musée du Louvre: G. Migeon, *Musée national du Louvre. Catalogue de la collection Paul Garnier*, Paris, 1917.
Paris, Musée du Petit-Palais: *Palais des Beaux-Arts de la Ville de Paris. Catalogue de la collection Tuck*, Paris 1931.
Paris, Musée national des techniques (CNAM): *Conservatoire national des Arts et Métiers. Catalogue du Musée. Section JB. Horlogerie*, Paris, 1949.
Toulouse, Musée Paul Dupuy: *Musée Paul Dupuy. Horlogerie et instruments de mesure du temps passé*, Toulouse, 1979.
Wuppertal, Uhrenmuseum: J. Abeler, *5000 Jahre Zeit-Messung*, Wuppertal, 1978.
Zurich, Musée de la mesure du temps: A. T. Beyer, *Antike Uhren. Das Museum der Zeitmessung*, Beyer Zurich, Munich, 1982.

Catalogues of collections

Chapuis, A., *Montres et émaux de Genève, Louis XIV, Louis XV, Louis XVI et Empire, collection H. Wilsdorf*, Lausanne, 1944.
Collection de montres et automates Maurice et Edouard Sandoz, Le Locle, Musée d'Horlogerie, 1976.
Dubois, P. *Collection archéologique du prince Soltykoff. Horlogerie, description et iconographie des instruments horaires au XVIe siècle*, Paris, 1858.
Gélis, E., *L'Horlogerie ancienne, histoire, décor et technique*, Paris, 1950. [Gélis collection].
Hubert, M., Banbery A., *Patek Philippe*, Geneva, Zurich, 1982.
Palustre, L., *Les Horloges et les montres*, in *La Collection Spitzer*, t. V, Paris 1892.
Parisi, B., *Catalogo descrittivo della collezione delle Piane di orologi da petto e da tasca*, Milan, 1954.
Pippa, L. *Orologi nel tempo da una raccolta*, Milan, 1966.
Williamson, G. C., *Catalogue of the collection of watches, the property of J. Pierpont Morgan (. . .)*, London, 1912. Reprinted, Paris, 1972.

Catalogues of exhibitions

1900, Paris, Exposition Universal, *Musée rétrospectif de la classe 96, Horlogerie*.
1921, Paris, Musée Galliera, *Décor moderne de l'horlogerie et de la bijouterie*.
1923, Paris, Musée Galliera, *Exposition de la verrerie et de l'émaillerie moderne*.
1923, Paris, Musée Galliera, *Centenaire d'Abraham-Louis Breguet*.
1926, Paris, Musée des Arts Décoratifs, *Exposition d'orfèvrerie française civile du XVIe au début du XVIIe siècle*.
1949, Geneva, Hôtel Métropole, *Montres et bijoux*.
1949, Paris, Musée du CNAM, *Chefs d'oeuvre de l'horlogerie*.
1954, Paris, CNAM, *Horloges et automates*.
1970, La Chaux-de-Fonds, Musée international d'horlogerie; Besançon, Musée de l'horlogerie, *Montres émaillées des XVIe et XVIIe siècles. Collections du Louvre et des Musées parisiens*.
1970, Rome, Institut suisse, *Montres du XVIe au XIXe siecle. Collections des Musées d'horlogerie Le Locle, La Chaux-de-Fonds*.
1974, New York, Fifth Avenue, Rolex, *Four Hundred Years of Watchmaking*.
1976, La Chaux-de-Fonds, Musée international d'horlogerie, *L'Oeuvre d'Abraham-Louis Breguet*.
1978, Geneva, Musée de l'horlogerie, *Montres de Genève*.
1979, Geneva, Musée de l'horlogerie, *Montres à Stackfreed*.
1980, Besançon, Musée des beaux-arts, *Collections horlogères en Franche-Comté*.
1980, Munich, Bayerisches Nationalmuseum; Washington, Smithsonian Institute, *The Clockwork Universe*.
1982, Saint-Omer, Musée de l'hôtel Sandelin; Evreux, Musée de l'ancien évêché, *La Montre de 1580 à 1930*.
1983, Geneva, Musée de l'horlogerie, *Montres françaises 1580–1680*.
1984, Brussels, Société générale de banque, *La Mesure du temps dans les collections belges*.

Archival documents

JULY 18, 1459

Payment to Jehan de Limbourg, "master clockworker," for five clocks delivered to the king, one of which is without counterweights.

Archives nationales, KK 51, fol. 97 r°–v°

A Jehan de Lyebourg, maistre ouvrier d'orloges demourant a Paris, la somme de IIIIXX XVI livres V sols tournois en LXX escuz d'or que le Roy nostredit sire lui a ordonné estre baillez et delivrez comptant, c'est assavoir pour la vente de cinq orloges, les quatre desquelz sont a cloche et contrepoix et l'autre n'est que demy orloge doré de fin or sans contrepoix, qu'il a fait prandre et acheter de lui qui les avoit aportees oudit moys de Juillet par ordonnance dudit sire dudit lieu de Paris a Razilly prex Chinon devers lui qui les avoit retenues et fait mettre devers soy, en LIII escuz d'or, LXXII livres XVII sols VI deniers tournois,
et pour la voiture desdites cinq orloges, aussi pour la despence qu'il avoit faicte l'espace de troys sepmaines a se jour audit lieu de Chinon en attendant son expedition, en XVII escuz, XXIII livres VII sols VI deniers tournois,
qui est pour lesdites deux parties ladite somme de IIIXXXVI livres V sols tournois payée audit Liebourg par sa quictance sur ce faicte le XXIIIe jour dudit moys de Juillet, l'an MCCCCLIX, cy rendue

IIIIXXXVI l. V s.t.

APRIL 4, 1480

Payment to Jehan de Paris, *"orlogeur,"* for a portable clock destined for the king's usage.

Archives nationales, KK 64, fol. 124 r°–v°

A Jehan de Paris, orlogeur, la somme de seize livres dix deniers tournois en dix escuz d'or a luy ordonnée par ledit seigneur oudit mois de mars pour une orloge ou il y a ung cadran et sonne les heures, garnies de tout ce qu'il luy appartient, laquelle ledit seigneur a fait prandre et achecter de luy pour porter avecques luy par tous les lieux ou il yra. Pour cecy, par vertu dudit roole du Roy et quictance dudit de Paris escripte le quart jour d'avril mil CCCC IIIIXX, cy rendue ladite somme de

XVI l. X d. t.

JUNE 5, 1481

Payment to Marin Guerier for the transportation of the king's clock.

Archives nationales, KK 64, fol. 138

Audit Marin Guerier la somme de cinquante solz tournois a luy ordonnée par ledit seigneur pour avoir porté sur son cheval l'orloge dudit seigneur durant dix jours entiers oudit mois de fevrier, qui est a la raison de cinq solz tournois par jour. Pour cecy, par vertu dudit roole du Roy et quictance dudit Guerir rendue sur la partie precedente, servant cy pour ladite somme de

L s. t.

JULY 9, 1523

Clocks belonging to Margaret of Austria (1480–1530), mentioned in the inventory of her palace in Malines, drawn on July 9, 1523.

Bibliothèque Nationale, département des manuscrits, cinq cents Colbert, 128

fol. 93 v°

Item ung reloge en manière d'une tour, le dedans de fer et la cloture d'argent, à six carrés et six pilliers à l'entour, dorez, ouvrez à jour dessus, en chief ung bouton esmaillé de bleu.
Item ung aultre petit reloge en manière d'une pomme avec une petite chaînette dorée, a ung anneau au bout de ladite chaînette.

fol. 98

Item ung reloge de leton doré, fait en manière de torrelle losainger, eschequeter à l'entour; et dessus est escript la devise de Madame et dessoubz le cadrant a ung escuson aux armes de Madite Dame.
Item ung reloge rond, de leton doré, le cadrant d'esmail bleu, au milieu d'icelluy une marguerite et de coste une penseée esmaillée.
Item ung aultre reloge rond, plus haultet que le precedent et non sy large, de leton doré, au milieu du cadrant il y a ung sainct Philibert esmaillé.
Item ung aultre petit reloge de leton doré rond, au milieu du cadrant il y a une pensée de bleu esmail.
Item ung aultre reloge ung peu moindre, le cadrant a XII raiz de soleil, et au milieu le nom de Jhesus esmaillé.

DECEMBER 9, 1534

Inventory of the estate of Spire de Pons, wife of Nicolas Morel, master clockmaker in Paris, living in the rue des Juifs and keeping a shop in the rue des Arcis. Expert witnesses: Jehan de Prelles and Loys François, master clockmakers.

Archives nationales (Minutier), étude XIX, liasse 265, fol. 5 v°–7 r°

Ensuyt la marchandise dudit mestier d'orlogeur.

Premierement, une monstre d'orloge de leston doré, garnye de son estuy aussi doré fermant a clef, prisée IX l. t.

Item deux estotz, l'un grant et l'autre petit, prisés ensemble C s. t.

Item une tenailles a viz, thenailles de boys, ung bigorneau, ung taceau carré et ung autre petit taceau carré, ung petit rond, troys tas a moulleurre, cinq cloches, quatre triboulletz de boys, une truyere servant a la forge et deux autres taceaux, ung autre bigorneau assis sur ung petit bloc de boys, une tenaillettes, troys compas, troys petiz marteletz, deux petites monstres d'orloges commancées rondes et plusieurs utancilles dudit mestier d'orlogeur, prisés ensemble X l. t.

Item six marteaulx de fer dont deux a maçon et les quatre servans audit mestier, une lampe de fonte de cuyvre garnye de sa cremillere de fer, une cueillier a fondre et autres plusieurs vieilles ferrailles, prisés ensemble X s. t.

Item une saulciere et ung esgoutouer, le tout d'estaing, prisés ensemble III s. IIII d. t.

Item une escabelle, ung muy et ung demy muy a gueulle bée et deux ruylles, prisés ensemble X s. t.

Item une enclume asserée garnye de son bloc, prisée X l. t.

Item une grande bigorne garnye de son bloc, ung tranchet et ung petit tas dessus ledit bloc, prisés ensemble C s. t.

Item deux gros marteaulx et quatre petiz dont l'un sans teste, troys tenailles dont une croche et deux droittes, troys ressors servans a fere monstres, et plusieurs vieilles ferrailles et ung petit moulleau my monté, prisés le tout ensemble XL s. t.

Item deux souffletz garniz d'un treteau, d'une croisée et trueillée, deux auges de pierre dont l'une grande et l'autre petite, prisés ensemble LX s. t.

Du jeudi ensuivant, X[e] jour desdits moys et an, en continuant a la perfection dudit inventaire, fut fait d'icelluy ce qu'il s'ensuit:

Premierement, en l'ouvrouer assis rue des Arciz, fut trouvé ung gros estot et ung petit estot de fer, prisés ensemble VI l. t.

Item ung petit bigorneau, ung autre moyen bigorneau, ung petit tasseau, cinq autres petitz tasseaux a moulleurres, une petite tenailles a viz, ung compas a teste, deux ryvouers, une tenaillettes, le tout de fer, une autre tenaille de boys a viz, prisés ensemble XX s. t.

Item deux douzaines de lymes, tant grans que petites, garniz de leurs manches, tant de boys que de bouys, prisées ensemble X s. t.

Item deux tables carrées taillées, deux corps de bouettes a six pans, quatre bouettes, deux corps de bouette, ung autre corps de bouette garny de sa caige et une autre vieille boette a six pans, une autre petite caige carrée, une autre caige d'orloge a sonner a six pans et une autre caige a fuzée, le tout de leton, prisés ensemble L s. t.

Item une layette de boys en laquelle y a quelque quantité de leton my taillés, six autres layettes, tant grans que petites, dont l'une servant a mectre cordes de boyau, une paire de cisailles de fer, ung rouet a deux roues imparfaict, deux paires de platynes, une douzaine et demye de roues non dantées pour servir a monstres d'orloge non sonnans comme pilliers servans a reveille matin, ung petit tour de boys, ung gros vieilhebrequin garny de son fust, ung autre fust de vielbrequin, deux petiz estuyz de monstres dorez, ung petit mousleau garny de son auge, une caige d'un reveil matin garny de roues dantées imparfait, ung marteau de fer a une teste et plusieurs rongneures de fer et ferrailles, tant petites que grandes, le tout prisé ensemble C s. t.

Item huit petiz tymbres de fonte, les ungs telz quelz, servans a my taille et six petites clochettes, prisés ensemble XV s. t.

Item deux selles rondes, chacune a troys piedz, une petite monstre de boys a quatre piedz, de deux piedz de long ou environ, sept aiz servans de tablettes de plusieurs longueurs, ung petit caque a gueulle bée, ung seau, une sourissiere, une ratissouere et ung petit billot de boys, prisés ensemble VII s. VI d. t.

JANUARY 27, 1547

Inventory of the estate of Barbe Nepveu, wife of Master Hillaire Josse, surnamed de La Chappelle, barrister in parliament, living in the rue de la Vannerie, under the signboard the Golden Door. Expert witnesses: Jehan Bourillon and Jacques Fieffé, master clockmakers in Paris.

Archives nationales (Minutier), étude LXXXVI, liasse 93, fol. 18 v°–20 r°

Le mardi premier jour de fevrier oudit an V^C^XLVI, en continuant a la perfection dudit inventaire fut inventorié ce qui s'ensuict

Et premierement dedans les aulmoires de la garde robbe de ladite premiere chambre dudit hostel de la Porte Dorée, fut trouvé la marchandise d'orlogerie et monstres prisées par lesdits Bourrillon et Fieffe

Et premierement une grosse orloge de deux piedz et demi en carré ou environ, prisée vingt cinq livres tournois, pour ce — XXV ll. t.

Item une aultre moyenne orloge qui remonste le contrepoix a une manielle, garnye de sa roue et de cadran, prisée treize livres dix solz tournois, pour ce — XIII ll. X s. t.

Item une aultre orloge plus moyenne sans cadran, imparfecte, garnye de ses contrepoix, prisée huit livres dix solz tournois, pour ce — VIII ll. X s. t.

Item une aultre moyenne orloge a appeaulx, sans cloche et contrepois, prisée neuf livres tournois, pour ce — IX ll. t.

Item une aultre orloge a chanterie, garnye de ses appeaulx, prisée unze livres cinq solz tournois, pour ce — XI ll. V s. t.

Item une aultre orloge a appeaulx plus moyennes que les precedantes, toute garnye – Reste que le cadran n'i est point –, prisée huit livres tournois, pour ce — VIII ll. t.

Item une aultre orloge d'un demi pied, garnye de sa cloche, en laquelle deffault la roue et pillier de cadran, prisée cent solz tournois, pour ce — C s. t.

Item une aultre petite orloge qui destant par le cadran le guichet de cuyvre, garnye de ses contrepoix et de son estuy, prisée neuf livres tournois, pour ce — IX ll. t.

Item une petite orloge d'une livre garnye de ce qui luy fault, prisée six livres quinze solz tournois, pour ce — VI ll. XV s. t.

Item une aultre orloge carrée, garnye de son cadran et cloche, prisée six livres tournois, pour ce — VI ll. t.

Item deux petites orloges imparfectes, prisées ensemble quarente cinq solz tournois, pour ce — XLV s. t.

Item ung reveille matin carré, prisé soixante solz tournois, pour ce — LX s. t.

Item deux petites monstres dorées, prisées ensemble cent dix solz tournois, pour ce — CX s. t.

Item une aultre petite monstre dorée, garnye de son estuy feré, prisée quatre livres dix solz tournois, pour ce — IIII ll. X s. t.

Item ung mouvement d'une monstre d'orloge ronde, garnye de son estuy de cuyr doré, prisée quarente cinq solz tournois, pour ce — XLV s. t.

Item deux tempestes monstées sur du bois, dont l'une garnye de sa cloche, prisées ensemble vingt solz tournois, pour ce — XX s. t.

Item une laiette garnye de plusieurs lymes servant a orlogeur, prisées ensemble quinze solz tournois, pour ce — XV s. t.

Item quatre marteaulx, ung compas, une paire de tenailles longues, une petite quantité de fil de fer et plusieurs petitz oustilz servant a orlogeur, prisez ensemble vingt solz tournois, pour ce — XX s. t.

Item plusieurs timbres et clochettes de mestail pesant ensemble vingt et huit livres, prisé la livre deux solz tournois, vallant ensemble audit pris cinquante six solz tournois, pour ce — LVI s. t.

Item ung rotissouere de fer a contrepois garny de roues, prisé quatre livres dix solz tournois, pour ce — IIII ll. X s. t.

NOVEMBER 19, 1569

Sale of a rock-crystal watch, ornamented with a pearl, rubies and diamonds, belonging to Yves de Brinon.

Archives nationales (Minutier), étude CVII, 8

Yves de Brinon, escuyer seigneur de Sardes et Guyencourt demeurant à Paris consent et accorde qu'il soyt baillé a Michel (?) Pluys marchant mercier demeurant à Paris, à ce présent, une monstre d'orloge enchassée en cristal garnye d'or par les garnitures avec rubis et diamens alentour et une perle au bout, estant entre les mains de madame Regnault femme de monsieur Regnault, advocat, que ladite dame Regnault tient en gage dudit seigneur de Guyencourt pour la somme de cinquante escuz. En payant ledit Pluye lesd. cinquante escuz a lad. dame, laquelle monstre icelluy Pluye a prins et prent pour la somme de C escuz sur et en déduction de plus grant somme que led. seigneur de Guyencourt doit au dit Pluye et laquelle en ce faisant led. Pluye sera tenu en tenir compte au dit seigneur de Guyencourt, et en ce faisant led. seigneur de Guyencourt en a quitté et déchargé, quitte et décharge lad. dame Regnault . . . l'an mil V[c] LXIX le samedi 19e jour de novembre.

1661

Watch and clock belonging to Cardinal Mazarin (1603–1661), mentioned in his death inventory.

Bibliothèque Nationale, département des manuscrits, mélanges Colbert, 75

fol. 6 v° et fol. 7 r°

Une grande monstre ronde sonnante a boette d'or sous email gravée dont le mouvement est de Macé orlogeur a Blois, prisée la somme de mil livres.

Une autre grande montre sonnante ronde a boette de leton doré dont le mouvement est de Gregoire a Blois, prisée la somme de trois cents livres tournois.

Une autre grande montre sonnante ronde dont le mouvement est de Pierre Le Roux a Blois a boette d'or sous email a la . . . (?) cadran esmaillé de vert, prisée la somme de mil livres tournois.

Une autre grande montre sonnante ronde dont le mouvement est de Jean Roux a Blois a boette d'or le cadran emaillé de vert, prisée pareille somme mil livres.

Item une autre montre sonnante ronde a boette d'argent dont le mouvement du nommé . . . beaussi (?) a Saumur, prisée la somme de cent cinquante livres.

fol. 9 r° et v°

Une orloge sonnante garnye d'or d'esmaux dapli coulleur Viollet, garny de quarante quatre petits diamants et de son estuy de velours rouge, prisé ensemble la somme de trois cents livres.

APRIL 28, 1724

Watches belonging to Philippe d'Orléans (1674–1723), mentioned in the inventory of his estate, drawn on March 10, 1724, and the following days. They carry the numbers 1239 to 1242 in the collection of "jewels appraised at their right value and without increase, by the said Tesnière, with Charles Delafrenaye and Philippe Legras, merchant jeweler-haberdashers in Paris, advising." April 28, 1724.

Archives nationales X[1A] 9161 and 9162, fol. 113

Item deux montres d'or a repetition dont une gravée faite par Gaudron, et l'autre unie a repetition sourde d'Angleterre, avec leurs chaines d'or, leurs crochets l'un de metail et l'autre d'acier doré, a une desd.[s] chaines est attaché deux cachets de pierre gravée a la montre d'Angleterre, prisées ensemble a juste valeur et sans crüe la somme de mil livres.

Item une montre faite par Martineau a repetition dans sa boëte d'or, garnie de plusieurs petites pierres de grenats et diamants avec une chaine d'or, dont les claviers sont d'acier et le dedans de lad. boëtte de cuivre doré, prisée a juste valeur et sans crüe, la somme de deux cens cinquante livres.

Item une montre d'Angleterre a deux boëttes d'or avec sa chaine d'or et un crochet d'acier, prisée a juste valeur et sans crüe la somme de deux cens cinquante livres.

Item quatre Montres avec des boettes de chagrin, les dedans d'or, dont deux a repetition avec leur chaîne d'or, a l'une desquelles est un crochet et l'autre sans crochet, prisées ensemble a juste valeur et sans crüe la somme de cinq cens livres.

Acknowledgments

I would like to pay my respects and express my gratitude to those who generously helped in the present work and who, according to their functions or inclinations, are also contributing in building the history of horology: Daniel Alcouffe, head curator at the Musée du Louvre; Pierre Imhof, president of the Musée international d'horlogerie; André Curtit, curator of the Musée international d'horlogerie; William Andrewes, curator of the Time Museum; Beresford Hutchinson, curator at the National Maritime Museum (Greenwich); Giuseppe Brusa, curator at the Museo Poldi Pezzoli; Adolphe Chapiro, president of the Association nationale des amateurs et collectionneurs d'horlogerie ancienne; Jean-Claude Sabrier, the horology expert at the Cour d'appel in Paris, and Jean-Claude Gendrot, Michel Journe, Yves Mongrolle, Roberto Panicali, Jean-Nérée Ronfort and Anthony Turner.

In the realization of this book, I have particularly appreciated the aid of Philippe Henrat, curator at the Archives nationales, whom I heartily thank. My sincere thanks go equally to Mme Grodecki, curator of the Minutier des Archives nationales; Catherine Join-Dierle, curator of the Musée du Petit-Palais; Françoise Vittu; Jurgen Abeler, curator of the Uhrenmuseum in Wuppertal; Richard Good, curator at the British Museum; and Jeremy Evans.

Finally, I wish to pay a respectful and grateful homage to Jacques Thuillier, professor of the Collège de France, who was the first to encourage me in undertaking this work.

Index

The page numbers in italic refer to the captions for illustrations.

Photographic credits

We wish to thank here the photographers who took the pictures, the museums, the collectors and the institutions for their courtesy in letting us use the photographs reproduced in the present work. The numerals correspond to the illustration numbers.

Bibliothèque Nationale, Paris 37, 38a, 38b, 38c, 73, 75, 77, 79, 100, 102, 103, 110, 117, 121, 126, 136, 138, 141, 144, 146
Bibliothèque Royale, Brussels 1
Bildarchiv Preussischer Kulturbesitz, Berlin 2
H. R. Bramaz, Kilchberg 180
British Museum, London 7 left, 7 right, 16 left, 16 right, 17, 71, 83, 90, 93, 95, 98 left, 98 right, 103, 154
J. E. Bulloz, Paris 9, 15, 133 top, 133 left, 133 right
Jean Dieuzaide, Toulouse 109
Marianne Haller, Sudstadt 18
Leo Hilber, Fribourg 4, 8 left, 8 right, 20 left, 20 right, 21 left, 21 right, 25, 34, 62a, 62b, 66a, 66b, 80, 97, 101, 106, 113a, 113b, 115, 150, 169, 172
Lord Chamberlain's Office, London 55
L. A. Mayer Memorial, Institute for Islamic Art, Jerusalem 61
The Metropolitan Museum of Art, New York 125
Musée d'art et d'histoire, Geneva 58
Musée des beaux-arts, Caen 36, 39
Musée de l'horlogerie et de l'émaillerie, Geneva 91
Musée international d'horlogerie, La Chaux-de-Fonds 107a, 107b, 107c, 107d, 179
Musée Paul Dupuy, Toulouse 145
Museo Poldi Pezzoli, Milan 114, 131, 137
Musées royaux d'art et d'histoire, Brussels 157, 165 left, 165 center, 165 right, 171 (photos: Bernard Daubersy)
National Galleries of Scotland, Edinburgh 160
National Maritime Museum, Greenwich 48, 51, 52
Patek Philippe, Geneva 67, 178
Photo Création SA, Le Locle 149, 168, 170, 174
Réunion des Musées nationaux, Paris 10, 72, 74, 76, 78, 84, 85, 86, 87, 88, 92, 105, 108a, 108b, 111, 117, 118a, 118b
Rijksmuseum, Amsterdam 104a, 104b
Staaliche Kunstsammlungen, Kassel 40, 130
Studio Lourmel, Photo Routhier, Paris 6, 11, 12, 13, 14, 19 left, 19 right, 22, 23 left, 23 right, 24, 27, 29, 35, 43, 44 left, 44 center, 44 right, 45 top, 45 bottom, 46, 49, 50, 63, 64a, 64b, 65, 68, 69, 70, 89a, 89b, 112, 120, 122 left, 122 right, 123 left, 123 right, 124, 132, 134, 135, 139 left, 139 right, 140 left, 140 right, 142, 143 left, 143 center, 143 right, 147 left, 147 right, 148 left, 148 center, 148 right, 151, 155, 156 top, 156 center, 156 bottom, 163, 167 left, 167 right, 173
The Time Museum, Rockford (Illinois) 3, 28, 60, 175, 176 top, 176 bottom (photos: Stephen Pitkin)
Eileen Tweedy, London 81 left, 81 right, 94, 96 left, 96 center, 96 right, 99 left, 99 right, 116, 127, 128 left, 128 right, 129 left, 129 center, 129 right, 152, 153, 158, 161, 162 left, 162 right
Uhrenmuseum, Wuppertal 5; Gerd Hensel photos: 164, 166
The Worshipful Company of Clockmakers, The Clock Room, Guildhall Library, London 26 left, 26 right, 54, 82, 159 (photos: Godfrey New Photographics Ltd, Sidcup, Kent)

Author's archives: 57a, 57b, 177

The drawings were reproduced from the following works:
32, 33, 41, 53, 56: Meis, Reinhard, *Les Montres de poche*, Fribourg, 1980;
30, 31, 59: Guye, Samuel, and Michel, Henri, *La Mesure du temps et de l'espace*, Fribourg, 1970;
47: Bassermann-Joran, Ernst von, and Bertele, Hans von, *Montres, horloges et pendules*, Fribourg, 1961.